Wildlife Search and Rescue

COMPANION WEBSITE

This book has a companion website:

www.wiley.com/go/dmytryk/wildlifeemergency

with Figures and Tables from the book for downloading

Wildlife Search and Rescue

A Guide for First Responders

Rebecca Dmytryk

WILEY-BLACKWELL

A John Wiley & Sons, Ltd., Publication

This edition first published 2012 © 2012 by John Wiley & Sons, Ltd

Wiley-Blackwell is an imprint of John Wiley & Sons, formed by the merger of Wiley's global Scientific, Technical and Medical business with Blackwell Publishing.

Registered office: John Wiley & Sons, Ltd, The Atrium, Southern Gate, Chichester, West Sussex, PO19 8SQ, UK

Editorial offices: 9600 Garsington Road, Oxford, OX4 2DQ, UK
The Atrium, Southern Gate, Chichester, West Sussex, PO19 8SQ, UK
111 River Street, Hoboken, NJ 07030-5774, USA

For details of our global editorial offices, for customer services and for information about how to apply for permission to reuse the copyright material in this book please see our website at www.wiley.com/wiley-blackwell.

The rights of the author to be identified as the author of this work has been asserted in accordance with the UK Copyright, Designs and Patents Act 1988.

All rights reserved. No part of this publication may be reproduced, stored in a retrieval system, or transmitted, in any form or by any means, electronic, mechanical, photocopying, recording or otherwise, except as permitted by the UK Copyright, Designs and Patents Act 1988, without the prior permission of the publisher.

Designations used by companies to distinguish their products are often claimed as trademarks. All brand names and product names used in this book are trade names, service marks, trademarks or registered trademarks of their respective owners. The publisher is not associated with any product or vendor mentioned in this book. This publication is designed to provide accurate and authoritative information in regard to the subject matter covered. It is sold on the understanding that the publisher is not engaged in rendering professional services. If professional advice or other expert assistance is required, the services of a competent professional should be sought.

A special thank you to Rebecca Duerr, DVM, for helpful comments on the wildlife first aid section of the book.

The star image on the cover is a registered service mark (Registration number: **3757480**).

Library of Congress Cataloging-in-Publication Data

Dmytryk, Rebecca.
  Wildlife search and rescue : a guide for first responders / by Rebecca Dmytryk.
    p. cm.
  Includes index.
  ISBN 978-0-470-65510-8 (cloth) – ISBN 978-0-470-65511-5 (paper)
  1. Wildlife rescue–United States–Handbooks, manuals, etc. 2. First responders–United States–Handbooks, manuals, etc. I. Title.
  QL83.2.D58 2012
  333.95′417–dc23
                                                                                    2011030370

A catalogue record for this book is available from the British Library.

Set in 10/13pt RotisSemiSans by Laserwords Private Limited, Chennai, India.
Printed and bound in Malaysia by Vivar Printing Sdn Bhd

First   2012

This book is dedicated to wild creatures and their wild lands.
Aho Mitakuye Oyasin (to all my relations)

This book is intended solely as a guide to appropriate procedures for response to emergencies involving wild animals based on the most current recommendations of responsible sources, yet, because of the uniqueness of emergencies, certain situations may require additional safety measures other than those described herein. Therefore, readers are cautioned to use their best judgment and adopt all safety precautions indicated by their activities. Additionally, this book is not meant to be used to advise anyone as to their legal authority to capture, possess, and treat wildlife. By following guidance contained herein the reader willingly assumes all risks in connection with such activities.

Contents

Foreword Jay Holcomb		xiii
Preface		xvii
1	*Overview of wildlife rescue*	1
2	*Characterizing wildlife search and rescue*	3
3	*Laws and regulations governing wildlife rescue in the USA*	7
4	*Code of practice*	10
5	*The components of wildlife search and rescue*	11
	Human safety	11
	Environmental hazards	11
	Human factor hazards	12
	Equipment hazards	13
	Health risks	13
	Zoonotic diseases	15
	Bacterial infections	15
	Fungal infections	18
	Viruses	19
	Parasites	19
	Personal protective equipment	22
	Protection from hazardous materials	24
	Basic safety and preparedness guidelines	26
	Operational risk management	27
	Outfitting	32
	The welfare of the animal	34
	Understanding stress	35
	Minimizing stress during rescue operations	38
	Potential for success	40
	The mindset of the hunter and the hunted	42
	The importance of natural history	42
	The fundamentals of the search	43
	The fundamentals of the capture	45

6	*Anatomy of a response team*	52
7	*Overview of wildlife capture equipment*	55
	The towel	55
	Herding boards	55
	Nets and netting	57
	The hoop net	57
	The open-ended hoop net	58
	The throw net	60
	Land seine	60
	Mechanical nets	60
	Active land seine	60
	The bow net and Q-net	61
	The whoosh net	62
	Driving, funnel, and walk-in traps	63
	The dho-gaza	66
	Drop traps	66
	Cage traps	68
	Projectile-powered nets	68
	Lures	69
	Catchpole	70
8	*Capture, handling, and confinement of wild birds*	71
	Techniques for capturing wild birds	71
	Enticing wild birds using lures	71
	The Bartos trap	76
	Snare-type traps	76
	Bal-chatri	76
	The phai trap	78
	Noose carpets	79
	The single snare	79
	Leg snare pole	84
	Swan hook	85
	Pit traps	85
	Mist nets	86
	On the water	86
	Floating gill nets	88
	Floating barriers and submersible pens	89
	Spotlighting	90

	Special circumstances and particular methods	92
	Hummingbirds	92
	Loons (Gaviiformes)	93
	Grebes (Podicipediformes)	94
	Rails and coots (Rallidae)	94
	Brown pelican (Pelecanus occidentalis)	95
	Cormorants (Phalacrocoracidae)	97
	Waders	97
	Alcids	98
	Birds trapped in structures	98
	Hummingbirds in skylights	99
	Window strikes	100
	Ducklings in a pool	100
	Birds entangled in fishing tackle	104
	Rodenticide poisoning	105
	Shot through with a projectile	105
	Glue traps	109
	Avian botulism	110
	Lead poisoning	111
	Domoic acid poisoning (DAP)	111
	"Sea slime"	113
	Oil and petroleum products	113
	Handling and restraint of wild birds	117
	Processing from nets and housing	124
	Short-term and temporary housing for wild birds	124
9	*Capture, handling, and confinement of land mammals*	130
	Techniques for capturing wild mammals	130
	Chemical immobilization	131
	Special circumstances and particular methods	131
	Small rodents	131
	Large rodents, porcupines, beaver	132
	Lagomorphs, rabbits and hares	132
	Xenarthrans, anteaters, armadillos	132
	Skunks	133
	Canids	133
	Deer	133
	Physical restraint of land mammals	134
	Bats	134

x Contents

      Small rodents   135
      Talpids, moles and relatives   136
      Squirrels   137
      Opossums   137
      Porcupines   137
      Lagomorphs, rabbits and hares   138
      Small and medium-sized carnvores   138
      Mustelids, badger, otter, weasels   139
      Skunks   139
      Coyotes and foxes   139
      Felids   140
   Processing mammals from nets and cages   140
   Temporary confinement of land mammals   142

10 *Capture and handling of reptiles and amphibians*   145

11 *Marine mammal rescue*   146
   Rescuing seals and sea lions   146
   Young seals and sea lions   147
      Hoop nets   149
      Modified open-ended hoop net   149
      Wraps, slings, and stretchers   151
      The towel wrap   151
      Flat webbing cargo net   152
      The floating net   153
   Physical restraint of seals and sea lions   153
   Confinement and transport of pinnipeds   153
   Cetaceans   156

12 *Basic wildlife first aid and stabilization*   158
   Performing a cursory physical examination   159
   Bleeding   161
   Dehydration   161
   Fluid therapy   162
      Oral fluid administration (mammals)   164
      Oral fluid administration (birds)   165
      Subcutaneous injections   168
   Treating hypothermia   170
   Treating Hyperthermia   171
   Basic wound care   172

Contents **xi**

Stabilizing fractures	173
Robert Jones bandage	174
Figure-eight wrap	174
Bird body wrap	176
The ball bandage	176
13 *Transporting wildlife*	177
14 *Field euthanasia*	179
15 *Life, liberty, and euthanasia*	182
16 *Rescuing baby birds and land mammals*	184
17 *Reuniting, re-nesting, and wild-fostering*	185
Returning altricial chicks to the wild	187
Returning precocial chicks to the wild	193
Returning baby mammals to the wild	195
Nutritional support	198
18 *Offering public service*	200
Public relations and the art of shapeshifting	202
Contracting with municipalities	203
Appendix 1 *Ready packs*	204
Appendix 2 *Wildlife observation form*	206
Appendix 3 *Wildlife trauma equipment and supplies*	207
Appendix 4 *Instructions for tying nooses*	209
Appendix 5 *Barn owl box plans and instructions*	211
Appendix 6 *Sample contract*	213
Further reading	220
Index	222

COMPANION WEBSITE

This book has a companion website:

www.wiley.com/go/dmytryk/wildlifeemergency

with Figures and Tables from the book for downloading

Foreword

Two nights ago I had dinner with a friend who lives only a mile away from my home in a rural mountainous area. It was about 10 p.m. when I headed home. As I turned onto a short stretch of two-lane highway I immediately saw eyes in my headlights. The eyes belonged to a terrified doe that had been hit by a car and was struggling to get up off the pavement. She was very alert. Her front legs seemed to function, but as she attempted to pull herself up I noticed that her back end was not functioning. So there she lay, on the side of the road, thrashing and attempting to drag her injured body into the brush.

I drove about 50 feet past her and parked the car. At the same time, another car stopped about 50 feet before her and a man stepped out. I grabbed one of the large spare towels that I keep in the back of my car and quickly walked up to the struggling animal. She was breathing heavily and was in a severe state of stress and panic. I gently wrapped the towel around her head, being sure to cover her eyes. The towel also provided cushioning for her face. I then firmly pushed on the back of her neck to simulate the pressure of a predator. I pressed her to the ground and she responded by lying still.

I instructed my new assistant to take my place, behind her back, and gently but firmly maintain pressure to her neck to keep her still so I could examine her injuries. I instructed him not to pet or console her, but to just hold her firmly in place.

The man was very concerned and upset and every time she moved he wanted to pet her, hold her, and console her. I kept gently reminding him to stay to her back to avoid her powerful legs and to keep pressure on her – not to cuddle her, as she would not understand his sentiments. To her, he was a predator. He understood and did very well at controlling his urge to handle her unnecessarily.

I discovered that the deer had a broken back, broken pelvis, or something equally critical for an adult deer. I explained to the man that a deer with any injury that rendered it unable to walk was basically a dead deer, as there was no feasible way to help a wild, struggling, near 200-pound adult deer recover from such injuries. Plus, she was bleeding from her anus and her nostrils, which implied internal bleeding.

It was clear that the most humane thing to do for this poor animal was to euthanize her as soon as possible. My new friend sadly, but bravely, agreed. We switched positions again so he could call the local police department while I kept the deer as calm as possible. What was interesting to me was the conversation I had with this man during the 15 minutes that we waited for an officer to arrive.

The man looked me right in the eyes with deep emotion and awe and said, "How do you know this stuff?"

"Know what?" I responded.

"Know that her head needed to be covered, and how you predicted she would respond to us, and how to handle all of this?"

"Well," I said, "I rehabilitate wildlife for a living, so I have an advantage. I understand the nature and natural history of deer. I used to raise them from infancy and I have cared for many adult deer. I know that deer are prey species with an acute sense of hearing and sight. The first thing that needs to be done when encountering an injured deer is to do what you can to block the senses that trigger panic. So an old towel is the perfect tool to do that, plus it acts as a cushion. She could still hear us, but she was not seeing us, so her desire to bolt was greatly reduced. Because she is a prey species, when we put pressure on her neck she responded to what she probably experienced as an attack. A lion would bite her neck and hold her while he killed her. We applied the same kind of pressure to keep her still.

I placed you on the back side of her because her thrashing legs are very powerful and most of the injuries sustained by people who help injured deer are from thrashing legs.

"It really worked. I am so impressed and happy that you came along," he said in amazement, and he thanked me profusely for helping the poor animal.

When the officer arrived we discussed the situation with her and she fully agreed that the best thing to do for the deer was to shoot her. The officer told us how much she hated her job at times like this.

I left the scene saddened, of course, but somehow renewed because I felt my actions had helped an animal, if only for a short time, and perhaps had a positive influence on a very concerned person who witnessed the end of a wild animal's life that he had played a role in helping. I imagined he, too, came away sad, but inspired.

As I drove home, I realized that my actions were automatic. When I saw the deer on the road, I knew exactly what to do, and it was more than just caring for the deer.

What the man didn't know was that as I was parking the car I was already making a plan. I called my friend who I'd just had dinner with and put her on alert in case I needed assistance. Without even thinking, I grabbed the towel, and made a mental review of what supplies I had in the car. I had rope, plastic bags, extra towels, etc.

I also knew that this man who had also stopped to help the deer could be emotional and might respond in a way that would put him and me in danger, so I considered the best way to use his help and keep us both safe while helping the animal. Then, as I walked toward the deer with towel in hand, I decided whom to call should the deer need to be dispatched. So, by the time I arrived at the animal, I already had a plan in my mind. I also knew that the plan could and should be altered according to changes in the circumstances.

I did all of this automatically and swiftly – I am a wildlife rehabilitator, a first responder and wildlife paramedic.

Much of what I know, I have learned the hard way – through trial and error, especially in my early days. I did not have a book or a manual to guide me and I am sure I made my share of mistakes. This book, *Wildlife Search and Rescue: A Guide for First Responders*, is truly a Godsend. It is not only very much needed, it is the first book of its kind that emphasizes the simple and basic principles, like the ones I used in helping the deer, that are key aspects of being a first responder to wildlife emergencies. Know the natural history of the animals you are working with. Know what stimulates each species, and once you understand that, then you can decide how to calm them. Know stress, and know not to be afraid of it. Understand its purpose, and use it to your advantage. Know basic medical techniques so that you can minimize injury and an animal's suffering. Above all, know that human safety comes first. You cannot help animals if you are injured and as a first responder you are also responsible for the health and safety of others around you.

These and many other fundamentals, which may seem obvious, but aren't in a time of crisis, are clearly outlined in this book. This valuable book's intent is to provide you with smart and proven techniques to help you to be responsible, safe, and effective when attempting to aid wild animals in peril. In essence, this book is just like the towel – it's another tool to help you reduce suffering and save lives. Read it, understand it, and apply it, and, like me, you will "know" how to react to a wildlife crisis situation when you encounter one.

Jay Holcomb
Executive Director
International Bird Rescue

Preface

For wildlife casualties and the people who find them, often the greatest challenge is locating someone with the necessary skills who is available to help. In the absence of specially trained personnel, animals' lives are lost. This may be the most significant cause of death among wildlife accident victims – one that has, for the most part, gone unnoticed, undocumented, and ignored.

It would be far better if we could say that overall response is substandard, but the issue is far worse – there are no standards. Just as it was for Emergency Medical Services (EMS) prior to the 1970s, currently there are no standards for provision of care for disabled native wildlife. Other than laws regulating the possession of wild species, there is no mandate for response and no direction on "best practices."

The rescue of a wild animal in distress is not simply grabbing a net, chasing it down, and placing it in a box. From the moment an animal is discovered, the action taken – the type of equipment used, the degree of handling, the method of confinement, and what initial aid is provided – can mean the difference between life and death. It is that simple.

If I were to contribute one single thing to the field of wildlife rehabilitation, it would be this concept: the rescue of wildlife casualties begins not at the wildlife hospital, but in the field, when and where the animals are first found. The most heroic, most skilled rehabilitation efforts will do no good if the animal doesn't make it to the treatment table. Therein lies the objective of this book.

This book is an attempt to lay a foundation for standards for wildlife search and rescue, offering examples of basic "best practices." It is meant to be a tool for existing animal rescue programs, to help them build or strengthen their capabilities. It is also intended for use by impromptu rescuers.

Image credits

Unless otherwise indicated illustrations and photographs are by Rebecca Dmytryk © 2011.

> Figure 46 Rendering from photograph by Bill Hilton, Operation RubyThroat.
> Thanks to Kane Brides for permission to use his photographs.
> Beer can gull by permission of International Bird Rescue.

Figure 58 Rendering from a photograph from the book *Standards and Medical Management for Captive Insectivorous Bats.* Courtesy of Bat World Sanctuary/batworld.org.

Figure 59 by permission of Wildlife Center of Virginia.

Figures 67 and 68 by permission of Peter Wallerstein, Marine Animal Rescue.

1 Overview of wildlife rescue

On any given day in the United States, it can be estimated that tens of thousands of wild birds and mammals suffer injuries directly attributable to humans. Causes of such anthropogenic injuries include motor vehicles, power lines, domestic dogs and cats, pesticide use, pollution and oil spills, errant fishing line, and intentional acts of cruelty.

During spring and summer months, when tree trimming and brush clearing is rampant, the number of wild animals that are displaced or injured can triple. Entire populations can be impacted through loss of habitat and natural disasters, increasing the number of potential wildlife victims to an even greater figure.

The estimated number of wildlife casualties is based on records of wild animals that have been reported or rescued – a journey that begins when an animal is first observed by someone who is willing to seek help. With luck, the *finder's* quest will lead them to a specially trained wildlife professional.

The rescue of wild animals requires a unique set of skills, considerably different from those used in handling domestic animals. Not only do wild animals behave differently, capturing and handling them can be dangerous, especially for the animal. Wild creatures perceive approach and handling by humans as a threat; most will flee or fight for their lives, even if it kills them. How the animal is handled and the quality of care it receives can mean life or death. Even if the animal is not critically wounded, inadequate housing, mishandling, and improper food can be fatal. It is therefore imperative that first responders receive specific training on proper methods of tending to wild animals in peril.

First and foremost, they must learn the natural history of the species they're going to encounter. This will help them locate individuals in distress and help them determine if an animal is behaving normally or needs to be rescued. Familiarity with the species will also help responders plan safe and successful capture strategies. Responders must also be trained and equipped to handle and confine wild animals without causing additional harm. If an animal requires immediate aid, first responders must be capable of providing basic life-saving emergency care – similar to human search and rescue personnel and paramedics. Wildlife search and rescue technician (WSART), wildlife paramedic, wildlife trauma specialist, wildlife EMT – these are relatively new terms being used to describe a

---

Wildlife Search and Rescue: A Guide for First Responders, First Edition. Rebecca Dmytryk.
© 2012 John Wiley & Sons, Ltd. Published 2012 by John Wiley & Sons, Ltd.

specialized division of animal rescue. Unfortunately, however, these unique and valuable service providers are absent in most communities.

In the United States, for example, of the hundreds of licensed wildlife rehabilitation centers, relatively few provide field service. Instead of sending a team of experts into the field, they rely on animals being brought to their doors. This often leaves the actual rescue of an injured or ill wild animal to the finder. Finders who are unable or unwilling to perform a rescue themselves will look for help, which can be a daunting experience.

In some regions, game wardens will assist with calls regarding disabled wildlife. In urban environments, the duty of responding to reportedly disabled wildlife is often assumed by municipal animal control agencies as part of their public service. In either case, unless these officials are extensively trained and equipped specifically for wildlife, they can do more harm than good.

In the absence of someone with the skills to find, identify, assess, and provide immediate aid to wild animals in distress, countless lives are lost. This often-overlooked issue may very well be the greatest dilemma faced by wildlife casualties and the people who find them.

2 Characterizing wildlife search and rescue

The term Wildlife Search and Rescue, or Wildlife SAR, is used to define the action taken on behalf of a wild animal in distress. It is the act of responding to a reportedly disabled animal, or an animal in immediate danger, providing for its immediate needs and, where appropriate, delivering it to definitive care where it may recover and be returned to the wild.

Whether a community is faced with thousands of injured animals during a disaster, or confronted with a few individuals here and there, responders should be adequately trained, equipped, and prepared to provide professional Wildlife Search and Rescue services.

Some rescues can be as simple as providing advice over the telephone, or placing a hatchling back into its nest. Others can be gruelingly complicated, like untangling the antlers of a buck from a barbed wire fence. In any event, a foundation of sound principles and a required level of training can ensure safer, more successful missions. Yet, to date, there are no standards for provision of care for wildlife casualties, similar to the state of pre-hospital care in the United States prior to the 1970s.

In 1966, the National Academy of Sciences (NAS) published a study entitled *Accidental Death and Disability: The Neglected Disease of Modern Society*. Often referred to as the White Paper, the report quantified the magnitude of death and injury caused by accidents in the United States, and explicitly outlined deficiencies in emergency medical care.

The paper revealed that morticians were providing at least half of the nation's ambulance services, with hearses being used to transport patients. It also noted that the majority of emergency responders had little to no training in first aid or life saving techniques. Because of this, it pointed out that an injured person had a better chance of surviving on a battlefield, with a trained armed forces medic, than on a US highway.

The report went on to make specific recommendations, among them the idea of having a single nationwide number to summon emergency assistance; this would later become the 9-1-1 system used today. The paper also called for standards for the provision of emergency medical care. This included training criteria for emergency responders and attention to ambulance design and function. This landmark publication was the impetus for major improvements from which the modern Emergency Medical Services (EMS) system was born.

Today, EMS is made up of a network of multifaceted service providers. Together, they have become the front line for public safety, providing aid from primary response to definitive care: for example, from the scene of an accident to a hospital. These first responders receive special training in rescue, stabilization, transportation, and advanced treatment of trauma and other medical emergencies.

To identify their unique service, an emblem emerged in 1973 known as the Star of Life. Each of the star's six points represents a stage of service: detection, reporting, response, on-scene care, care in transit, and transfer to definitive care. The response to a reportedly injured animal shares these phases of action.

A wildlife rescue begins when a person first discovers a wild animal they believe is in distress. The second stage is their call for assistance. Ideally, they will reach a wildlife Call Taker, similar to a 9-1-1 dispatcher, who provides an initial assessment of the situation and advises the caller on what to do. Once rescue personnel arrive on scene they evaluate the animal's condition further, and tend to its immediate needs. If it requires additional treatment, the animal is transported to definitive care – a wildlife hospital or wildlife rehabilitation center.

The term "wildlife rehabilitation" means the professional nursing of sick, injured, or orphaned wild animals with the intent of returning them to their natural habitat. Most wildlife rehabilitators are licensed and have years of training and guidance. Rehabilitation centers can range from large, high-tech sprawling compounds to modest, home-based operations.

During convalescence, animals must be housed and cared for so they retain their wildness and their natural fear of humans. Before an animal can be released it must be free of disease and show that it has a reasonable chance of survival in the wild. It must be capable of obtaining appropriate food and shelter, and have the ability to escape predation.

Prior to release, some rehabilitation programs outfit their patients with an identifying mark. Mammals may be fitted with a tag, often on the ear, whereas birds are typically ringed with a metal band on one leg. This marking helps track the success of freed patients. Despite numerous documented cases of long-term survival, wildlife rehabilitation efforts are, at times, called into question.

From a wildlife conservation perspective, the rescue and rehabilitation of individual animals receives occasional criticism for being a waste of time and resources. Unless the process involves an endangered species or a large population, wildlife rehabilitation is largely discounted.

What critics often fail to realize is that wildlife rescuers and rehabilitators keep their skills honed through everyday practice with more common species. When it comes time to treat an endangered animal or large numbers of patients, they are able to meet the challenge. Also under-recognized is the valuable role wildlife rescue organizations play in their communities and in spreading conservation values.

Wildlife rescue and rehabilitation programs provide a valuable public service to the communities they serve. By being available to assist with found wildlife, these programs reduce the public's handling and possession of wild animals, increasing public safety.

As first responders, receiving initial reports from the public about injured or ill animals, rescue organizations are often the first to notice trends in wild populations: for example, the effects of toxins, changes in the environment, or disease outbreaks. Their ability to alert authorities early on can reduce the overall number of animals impacted.

Wildlife rescue organizations also inspire environmental stewardship and influence the public's perception of wildlife positively. Simply through their existence these programs foster the idea that wild animals have value and that individual animals deserve to be treated for their ailments. Additionally, these programs usually offer the means for community members to connect with wildlife in a profound way, helping to broaden their understanding and appreciation of wild animals and their habitats. Connections might be sparked by a newspaper story, a classroom presentation, volunteer opportunities, or knowing that efforts to save a wild creature are not in vain – that someone else, an entire team of dedicated specialists, is there to help. Whatever the case, these connections are considerable, even life-changing.

For wildlife search and rescue personnel, the shared desire to help a wild animal is the common thread – the foundation upon which to develop a relationship with the finder. Sometimes it helps to understand a person's perspective. For some, the desire to help a wild animal will stem from a sense of duty, or obligation, especially if they feel responsible for the animal's mishap. For others, it could be transference of goodwill, or an act of compassion, or empathy. For others still, it might be a feeling of moral obligation to treat living things with respect, honoring an animal's intrinsic value as an individual and its right to live.

This compassion for the individual is depicted in an essay, entitled *The Star Thrower*, by anthropologist and author Loren Eisley (1907–77). It is a story that is often shared among rescuers.

In short, one morning, a gentleman is walking a long stretch of storm-ravaged beach where thousands of starfish lie grounded. He notices a human figure in the distance, farther down the beach. The person appears to be dancing. As he approaches he sees that it is not a dancer but a young man gently tossing starfish, one by one, out to sea. Confronting the young man he asks why he is throwing starfish into the ocean. The young man explains that if he doesn't get them back into the water soon they will die. The gentleman responds with the notion that the young man cannot possibly make a difference, not with the miles of coastline and thousands of dying starfish. The young man bends down, picks up another starfish and casts it beyond the breaking waves, replying, "It made a difference for that one."

The story of the Star Thrower exemplifies the power of the individual – the individual animal and the individual person. Coming face to face with an animal in distress is a very powerful, pivotal moment. The subsequent action taken by the person speaks volumes about their personal values and possibly those of the culture or society in which they live.

In general, providing care to injured, ill, or orphaned wild animals is considered the right thing to do. Programs that offer rescue and rehabilitative care for wildlife exist because it is what the people of a town or city or country want. Essentially, these programs are defined and perpetuated by the people. Oversight of the programs, however, is commonly assumed by government entities. Chapter 3 provides a summary of authorities and regulations that pertain to wildlife rescue and rehabilitation.

3 Laws and regulations governing wildlife rescue in the USA

In North America, the beginnings of wildlife protection can be traced back centuries to English Common Law, where the King held all wild game in "sacred trust" for the people. Colonialists brought with them, and followed, this basic principle.

Later, during the Industrial Revolution, a burgeoning urban population placed excessive demands on natural resources and wild game was hunted at unsustainable levels. It was around that same time that sport hunting became a popular leisure activity among the urban upper class. Conflicts developed between these two hunting groups, resulting in the establishment of the first conservation policies.

In 1842, a Supreme Court ruling reaffirmed the principle that no one owns wildlife, and that these natural resources are to be held in trust by government for the benefit of its people. This court ruling, combined with the advocacy of the sport hunters, resulted in the Public Trust Doctrine.

This movement toward wildlife conservation inspired treaties between Canada and the United States and the creation of a unique set of principles called the North American Model of Wildlife Conservation, emphasizing the importance of protecting wildlife species and habitats through sound science and active management. The model is composed of seven concepts: wildlife as public trust resources; elimination of markets for game; allocation of wildlife by law; wildlife should only be killed for a legitimate purpose; wildlife is considered an international resource; science is the proper tool for discharge of wildlife policy; democracy of hunting.

In the mid to late 1800s, with science rather than partisanship the tool by which to manage wildlife, branches of government were created to study the nation's fish and wildlife resources. These studies quickly revealed that many species had been hunted to near extinction. Scientists, conservationist groups, and, once again, sportsmen were quick to lobby Congress for greater protection of fish and wildlife. President Theodore Roosevelt was among those to spearhead this conservation movement. It is from those conservation efforts and first branches of government that contemporary natural resource management emerged.

Today, in the USA, wildlife is still considered a natural resource to be held in trust by government for the benefit of the people. The main trustee agencies are the United States Fish and Wildlife Service (FWS), National Oceanic and Atmospheric Administration's (NOAA's) National Marine Fisheries Service (NMFS), and individual state natural resource commissions. As trustees, charged with managing and protecting wildlife, these agencies also regulate wildlife rescue and rehabilitation.

On the state level, while private possession of native wildlife is prohibited, a majority of the states do allow rehabilitation of wildlife through a permitting process. Most permits indicate the species approved for rehabilitation, selectively restricting possession of certain species considered too dangerous or that could spread disease. These limitations can also apply to the rescue and transport of wild animals, which, in most states, requires a separate permit.

California, however, is one of the few states that allow rescuers, including the general public, to temporarily possess a native wild animal so long as it is relinquished to the appropriate authorities within 48 hours. This does not include possession of big game such as elk, pig, cougar, or bear, or threatened or endangered species. Nor does it supersede the Migratory Bird Treaty Act.

The Migratory Bird Treaty Act is legislation implementing conventions for international protection of nearly all bird species and is administrated by the FWS, a bureau of the Department of the Interior. Under this statute it is illegal for anyone to, among other things, pursue, capture, or possess any migratory bird, dead or alive, without a permit. This includes rescue efforts and rehabilitative care, with an exception for licensed veterinarians.

Veterinarians may temporarily possess, stabilize, or euthanize a sick or injured migratory bird without a migratory bird permit so long as they transfer the bird to a federally permitted rehabilitator within 24 hours after its condition has stabilized. In addition, veterinarians are required to keep detailed records for up to five years for all migratory birds that die or are euthanized under their care. Should they receive a threatened or endangered species, they must notify the agency immediately.

Threatened and endangered species, and their ecosystems, are protected under the Endangered Species Act (ESA), established in 1973. The NMFS and the FWS share in its implementation. In general, the FWS manages land and freshwater species while the NMFS manages marine species, including marine mammals through administration of the Marine Mammal Protection Act.

Under the Marine Mammal Protection Act it is unlawful for anyone to harass or otherwise disturb, or allow disturbance of, a marine mammal by causing disruption to its normal behavior. However, through authorization from the NMFS and under conditions set forth in the Marine Mammal Protection Act these animals may be rescued and rehabilitated. This usually requires a Letter of Authorization; however, there is an exception.

Under the Code of Federal Regulations (CFR) there is a provision that allows government officials and authorized non-government entities to rescue and transport marine mammals for rehabilitation purposes. Statute 216.22 can be found under CFR Title 50, Chapter II, Subchapter C, Part 216, Subpart c.

What is important about this rule is that it provides a way for responders to aid marine mammals without a Letter of Authorization from the NMFS. Instead, through a contractual agreement with their municipality, wildlife rescuers can become authorized to recover marine mammals as officials of that agency. The first non-government entity to make use of this statute was Wildlife Emergency Response through a contractual relationship with the City of Malibu, California, in 1996.

In addition to being governed by state and federal regulations, wildlife rescue and rehabilitation activities may also be regulated on a municipal level through local ordinances. Local ordinances are often more stringent than state or federal law, and can vary greatly, even between neighboring communities.

Because regulations may vary or may change from time to time, it is important for responders to be familiar with current laws that govern their efforts. They must also consider any special circumstances. For instance, if presented with a wild animal that is a victim of a crime, law enforcement officials could require additional documentation and special handling if the animal is considered evidence. Casualties of an oil spill are treated similarly.

Interestingly, despite the numerous rules, regulations, doctrines, and treaties meant to protect wildlife, no government agency is responsible for the rescue and rehabilitation of injured or orphaned native wildlife. Historically, this important work has been taken up by non-government entities and individuals.

Early on, the pioneers of wildlife rehabilitation learned how to successfully care for debilitated wildlife through trial and error. Today, their knowledge, and that of other rehabilitation experts, can be found in a number of books and published papers. One such publication, *Minimum Standards for Wildlife Rehabilitation*, is in its fourth edition. This manual has become of great significance over the years, gaining popularity among regulatory agencies as a means of evaluating wildlife rehabilitation efforts. Unfortunately, though, while the compendium details techniques for caring for injured, diseased, and orphaned wildlife, there is little guidance on rescue procedures. The following chapter provides a set of basic principles relating to wildlife rescue operations.

4 Code of practice

1. In the course of responding to and rescuing a wild animal, human safety will take precedence; no human life will knowingly be placed in jeopardy to save an animal.
2. Wildlife rescues will be performed in accordance with all applicable laws.
3. Every effort will be made to reduce and minimize pain, stress, and suffering of an animal during capture, handling, and transport.
4. Documentation of the animal or the rescue attempt, whether through photography, video, or audio recordings, will be done in such a way that does not subject the animal to any additional or unnecessary handling or delay its care.
5. Capture strategies, handling techniques, tools, or methods of confinement that pose a significant risk to an animal's life shall be avoided.
6. The administration of advanced medical care will be provided under the direction of a licensed veterinarian.
7. Terminally injured or ill animals will receive the appropriate immediate care governed by the organization's written euthanasia policy.
8. When presented with an otherwise healthy but displaced young wild animal still requiring parental care or guidance, all options will be considered and every effort will be made to keep them with wild parents.
9. When returning animals to the wild, every effort will be made to return them to their home territory.

Wildlife Search and Rescue: A Guide for First Responders, First Edition. Rebecca Dmytryk.
© 2012 John Wiley & Sons, Ltd. Published 2012 by John Wiley & Sons, Ltd.

5 The components of wildlife search and rescue

Wildlife search and rescue operations can be separated out into three major components: human safety, the welfare of the animal, and potential for success.

Human safety must be a top priority and the prime directive of each and every rescue mission. No person's life shall knowingly be placed in jeopardy to save an imperiled animal. The second most important thing to consider is the welfare of the animal. Before action is taken on the animal's behalf, its safety and best interest must be thoroughly considered. The third component is the potential for success – success meaning no person was injured and no property damaged, no animal was harmed, and the proper action was taken on behalf of the wild creature.

Human safety

When responding to an incident involving wildlife, the entity or organization that takes the lead and directs search and rescue operations is ultimately accountable for overall safety. However, it is essential that each rescuer involved be personally responsible for their own safety and the safety of those around them.

The concern for human safety must extend to all persons involved, or who may become involved, including bystanders and other members of the public. This serves not only the people but the animal as well. If a human is injured during a rescue, attention will shift to the person and away from the animal.

One of the first steps in keeping safe is being able to identify hazards. A hazard is any real or potential condition that could cause injury or death, or damage to property. Risk, on the other hand, is the potential for that loss to occur based on an evaluation of associated factors.

Environmental hazards

Handling wild animals can be dangerous but often, during search and rescue missions, the surrounding environment can pose an even greater threat to human safety.

Wildlife Search and Rescue: A Guide for First Responders, First Edition. Rebecca Dmytryk.
© 2012 John Wiley & Sons, Ltd. Published 2012 by John Wiley & Sons, Ltd.

When evaluating a situation, rescuers will want to look at how environmental conditions, such as the weather, are going to influence the mission and rescue personnel. Rain, wind, and lightning can produce hazardous conditions, as can high temperatures and humidity. If working where there is a marine influence, rescuers will want to consider the tides and surf. The terrain, too, must be evaluated for treacherous conditions such as slippery, moss-covered rocks, uneven ground, deep mud, or steep cliffs. Deep or swift moving water can also be dangerous.

Rescue workers must also look out for *biological hazards*, such as a region's potentially harmful flora and fauna. These include venomous snakes and spiders, toxic plants, parasites, microbes, and both wild and domestic animals.

Urban environments present a unique set of environmental hazards. They can be grouped into two categories, technologic and social. *Technologic hazards* are things that are man-made, such as structures, machinery, vehicles, electricity, chemical toxins, and pollutants. *Social hazards* relate to human society and human behavior, such as cultural differences, language barriers, and human activity, including crime.

Human factor hazards

In addition to assessing risks posed by the environment, personnel, too, must be considered for factors that could influence safety. These can be referred to as *human factor hazards*.

It is important that rescuers be suited for the rescue. They must be capable of carrying out assigned duties safely, without harming themselves or others. Therefore, in preparing for a rescue, in assembling teams and delegating positions, individuals should be evaluated for their physical capabilities, their skill, and their character. The key here is for potential rescuers to also evaluate themselves and acknowledge when a mission or task is beyond their level of competency or ability.

Physically, a rescuer must be fit for the job. Certain assignments might require excellent vision, while others might demand speed and endurance, or brute strength. Supervisors must take care not to assign jobs that are beyond a person's physical abilities as this can place the animal and other people in jeopardy. When evaluating rescue personnel, their state of health must also be considered: illness or fatigue can compromise performance and impair judgment.

In addition to being physically capable, rescue personnel must have the ability to carry out their assigned duties safely and effectively. In designating roles, it is important that the demands of the mission and appointed responsibilities do not exceed the rescuer's level of skill and training.

Rescuers must also be able to identify and comprehend elements that are critical to their safety and that of their teammates. This is often referred to as situational awareness, and it is key to safe and successful rescue missions. It means being sharply attuned to what is going on all around you.

Lastly, an individual's character must be considered. Every rescue mission will be unique and there will be certain character traits that will be more advantageous than others, depending on the situation, and then there are certain character traits that are not acceptable for wildlife rescue work. Extreme fear, carelessness, reckless behavior, poor judgment, complacency, excessive motivation, and machismo are characteristics that should be avoided. In contrast, individuals who are team players, perform well under stress, possess good communication skills, are resourceful, observant, and adept at problem solving can make for excellent wildlife rescuers.

Equipment hazards

While unforeseeable equipment failures can happen, many accidents involving equipment and machinery can be traced back to human error. The risk of mishaps can be lessened through proper and adequate training of personnel involved in the operation, transport, and maintenance of equipment. Such training should encourage awareness, concern, and prudence about hazardous situations. It should also include instruction on proper use of safety gear, as even protective equipment can become hazardous under certain conditions. For example, there have been instances where people wearing chest-high waders have slipped into deep water and drowned. Once the waders take on water they become nearly impossible to remove, even for the strongest of swimmers.

When a rescue mission involves the use of potentially hazardous equipment or machinery, the operation's safety briefing should include an overview of the associated risks so that all personnel are aware of the equipment hazards. Being knowledgeable about and prepared for potential accidents is the first step in staying safe.

Health risks

It is imperative that rescue workers be aware of the potential health risks associated with their activities. Rescue operations may call for personnel to work in severe conditions for extended periods of time. Environmental conditions that pose health risks include prolonged exposure to the sun, extreme temperatures, and humidity. Those who are not adequately prepared or fail to follow precautionary policies can suffer dehydration, heat stroke, or severe sunburn.

Overexposure to the sun can be prevented by wearing protective clothing and by using adequate sunscreen. Unfortunately, many of the common brands of sunscreen contain ingredients that are either known or suspected carcinogens, hormone disrupters, or toxic if ingested. These include the physical-barrier type sun blockers such as titanium dioxide and zinc oxide. In recent years titanium dioxide was reclassified as a carcinogen. Zinc oxide, by itself, is considered non-toxic and relatively safe; however, there is growing concern over the safety of the nano-scale particles used to make it more transparent.

Another hazard to be aware of is overheating – or hyperthermia. High temperatures, prolonged exposure to the sun, humidity, and physical exertion can contribute to heat injuries. There are three levels of heat injury: heat stress, heat exhaustion, and heat stroke.

Symptoms of heat stress, the mildest of the three levels, include sweating and headaches. Symptoms of heat exhaustion, the next level of heat illness, include clammy moist skin, heavy sweating, fatigue, dizziness, headache, nausea, and muscle cramps. The most severe condition, heat stroke, is a life-threatening disorder. As a rule, heat stroke is characterized by hot and dry skin; however, the victim's skin may be moist from previous sweating. Other symptoms of heat stroke include chills, dizziness, slurred speech, confusion, hallucinations, and coma.

To treat heat-related injuries, the person should be moved to a shaded, cool area immediately. Removing their outer layer of clothing will help them cool. If they are conscious and alert they may be offered sips of cold water. If their symptoms are severe, it may be necessary to soak them in cool water to bring down their temperature. Heat injuries can be serious and should receive medical attention without delay.

To avoid heat injuries, personnel should adhere to work/rest cycles that might incorporate break periods or limit work to the cooler periods of the day. They should also wear loose-fitting, breathable, light-colored clothing and make sure to drink plenty of water to stay hydrated.

As a rule, thirst is one of the first signs of dehydration. Therefore, it is important to drink enough water that you never become thirsty. The amount to consume over a period of time is also important. Instead of drinking an entire bottle of water all at once, drinking smaller amounts more often will give the body a steady supply of fluids. An example would be to drink 6 ounces (175 ml) every 20 minutes.

To achieve and maintain a good level of hydration, avoid carbonated drinks, caffeine, and alcoholic beverages. Sports drinks may contain beneficial electrolytes, but most commercial brands contain excessive amounts of sugar or sugar substitutes, which can be problematic in large quantities. A packet of powdered electrolytes added to a bottle of water is a healthier option.

A simple way to check hydration is by urine color. Clear to very faint yellow indicates that you are hydrated; anything darker is a sign that you are probably dehydrated and need to take on fluids.

Zoonotic diseases

Zoonoses are infectious diseases that can be transmitted from animals to humans. Zoonotic diseases are occupational hazards faced by those who work with or around animals. The illnesses can be classified as bacterial, fungal, viral, protozoal or parasitic infections. Pathogens are transmitted through direct contact, which includes: penetration and absorption through the skin or mucous membranes; inhalation; ingestion; and through a vector, such as a tick or mosquito. Some zoonotic diseases can be fairly innocuous, while others, such as rabies, can be fatal. The following is a brief look at some of the most common zoonotic diseases and their modes of transmission.

Bacterial infections

Anthrax is a disease caused by *Bacillus anthracis*. The spores of this bacterium have a hard exterior, allowing them to survive in the environment for years until conditions are right for them to grow. The disease is usually found in grazing animals such as sheep and cattle, which ingest the spores from the dirt. The spores of anthrax may also be present on hides and skins. In humans, anthrax can occur in three types, depending on how it entered the body. The cutaneous form of anthrax is where the spores enter the body through a cut or by an insect vector. It can produce a small, red, raised, insect-bite-like spot that turns into a sore with a thick black scab. It can be fatal if left untreated, to about one in five. Inhalation, a rarer form of the disease, is where the spores are breathed into the lungs. The gastro-intestinal form is usually contracted when a person consumes undercooked meat of an animal that had anthrax.

Brucellosis is a contagious infection caused by the *Brucella* bacterium. There are several species of *Brucella* commonly associated with a specific animal host. *Brucella* is found in domestic and wild animals. It can be found in deer, raccoons, foxes, and other animals, including marine mammals. Transmission is through contact with body fluid or tissues of an infected animal, or ingestion of contaminated milk. Symptoms include fever and aching joints and muscles. Doxycycline and rifampin have been used to successfully treat infection.

Ehrlichiosis ("air-lick-ee-oh-sus") is the general name for several related tick-borne bacterial diseases caused by the recently discovered organisms

Ehrlichia chaffeensis, Ehrlichia ewingii, and *Anaplasma phagocytophilum* (formerly *Ehrlichia phagocytophilum*).

Studies indicate that only Ixodidae (hard ticks) are associated with ehrlichiae. Ticks become infected after feeding on the blood of an infected animal during their larval or nymphal stage. Hosts include deer, elk, and wild rodents. Ehrlichiosis can be a severe illness; *E. chaffeensis* causes monocytic ehrlichiosis (HME), where the bacteria infect leukocytes, invading white blood cells called "monocytes." Initial symptoms include fever, headache, malaise, and muscle aches. Patients have been treated successfully with doxycycline.

Erysipelothrix rhusiopathiae is the cause of a bacterial infection also known as fish-handler's finger. The microorganism is widespread in nature and can infect cattle, sheep, horses, swine, fish, shrimp, crab, and birds. The gram-positive bacteria are transmitted through breaks in the skin, causing cutaneous infection characterized by a unique, raised, purple-reddish lesion accompanied by burning and itching. The infection may move deeper into the body, developing into the septicemic form of the disease. *E. rhusiopathiae* infection has been associated with septic polyarthritis, endocarditis, brain infarctions, and chronic meningitis. *Erysipelothrix* has proven to be susceptible to penicillins and cephalosporins.

Leptospirosis is a disease caused by a type of spirochete bacterium shed in urine. This disease is found in both wild and domestic animals. Leptospirosis is transmissible to humans through ingestion of urine, or contaminated food or water, and through absorption and inhalation. Symptoms may vary from vomiting to kidney failure. Leptospira are susceptible to penicillin antibiotics.

Lyme disease is transmitted by ticks. Early symptoms include an expanding red ring around the tick bite. Additional signs of infection include fatigue, recurring fever, vertigo, muscle and joint pain, headache, and enlarged lymph nodes. Days or even months later, more severe symptoms may develop, including chronic arthritis and neurological issues.

Seal finger is an infection attributed to a microorganism from the genus *Mycoplasma*. Mycoplasmas are simple, one-celled organisms lacking a rigid cell wall structure. Strains found in pinnipeds can cause the condition known as seal finger, a zoonotic infection transmitted through bites or through broken skin when handling seals or their carcasses. Symptoms include painful swelling and inflammation of the infected area, with or without destructive joint involvement. The skin around the wound can become taut and shiny. *Mycoplasma* infections have shown resistance to penicillin and erythromycin antibiotics, but have been treated successfully with tetracyclines.

Plague is a disease caused by the bacterium *Yersinia pestis*. In nature, rodents are the natural reservoir of this disease. It is usually their fleas that transmit the bacteria to animals, including domestic cats and humans. However, the bacteria can also be transmitted through body fluids from an infected animal, entering through breaks in the skin or contact with mucous membranes. Plague can also be contracted through inhalation from a cough or sneeze of an infected animal. There are three forms of plague: bubonic, characterized by swollen lymph glands; pneumonic; and septicemic. The flu-like symptoms associated with the disease include fever, chills, nausea, headache, and myalgia. Septicemic and pneumonic plague can be fatal if not treated in the first 24 hours of illness.

Rickettsial infections, such as typhus and Rocky Mountain spotted fever, are caused by infection by small, oblique, intracellular proteobacteria. The bacteria are transmitted through arthropod vectors.

Murine typhus (*Rickettsia typhi*) is present in the feces of infected fleas. It is transmitted to humans through abraded skin, through the bite of the flea, or when a person scratches the bite wound, inoculating the bacteria into their skin. Symptoms include fever, chills, headache, and myalgia. A rash may develop on the trunk and spread to the extremities, unlike in Rocky Mountain spotted fever.

Rickettsi rickettsii, the causative agent of Rocky Mountain spotted fever, lies dormant inside a tick until activated by a warm blood meal. It's then released into the saliva of the tick and transmitted through a bite. Therefore, relatively long exposure to an infected tick (24–48 hours) must occur before the organism can transfer to the new host. Symptoms in humans include fever, chills, and headache. A rash usually appears, beginning with the hands and feet, commonly on the palms and soles, spreading centripetally towards the trunk.

Rickettsia akari (Rickettsialpox) occurs in mice. The vector is the mouse mite. Rickettsialpox is considered a mild disease that begins with the formation of a papule, which ulcerates and quickly scabs over. The second set of symptoms may take 7–24 days to appear. These include fever, chills, and muscle pain followed, 2–3 days later, with a generalized rash. The pox eruptions heal within 2–3 weeks without scarring.

Q fever is a disease caused by the rickettsia *Coxiella burnetii*. It is endemic worldwide, found in domestic animals, including cattle, sheep, and goats, but has also been found in wild mammals, birds, and arthropods. *C. burnetii* is shed in spore-like form in the body fluids of infected animals. It can be found at high levels in amniotic fluids and the placenta. In the environment, the organism is resistant to heat, drying, and disinfecting agents, allowing it to survive for long periods. Humans are normally infected through inhalation of airborne material or from tick bites.

Q fever is characterized by an infection that can persist for more than 6 months, yet only half the people infected with *C. burnetii* will exhibit symptoms. Clinical signs of infection include high fever, chills, sweats, severe headache, malaise, non-productive cough, vomiting, and diarrhea. Up to half of those showing severe symptoms will likely develop pneumonia, and some may develop hepatitis. A more serious complication of chronic Q fever is endocarditis. Patients with acute Q fever may not develop the chronic form for years – as many as 20 years after infection. Doxycycline is the treatment of choice for acute Q fever.

Salmonellosis is transmitted through ingestion, by eating or drinking contaminated food or water, or fecal – oral contamination. Symptoms include vomiting and diarrhea. Personal hygiene and protective clothing will reduce exposure.

Tularemia, common in rabbits and rodents, can be transmitted through contact with infected tissue, ingestion of infected meat, aerosolized body fluids, or the bite of an infected insect vector. Symptoms include ulceration at the site of infection where the organism entered, and enlarged lymph nodes.

Fungal infections

Aspergillosis is caused by the ubiquitous fungus *Aspirgillus fumigatus*. Most healthy animals, including humans, will have a natural resistance to infection. However, the stress of captivity in wild birds can cause them to become vulnerable. Most susceptible are aquatic birds and raptors. Wild birds should never be housed on straw or wood shavings where the fungus can thrive. Indoor housing must be well ventilated with a minimum of 12 air exchanges per hour in rooms housing susceptible species. Transmission to humans is through inhalation of spores. Persons who are debilitated or immunosuppressed will be more at risk of infection.

Histoplasmosis is caused by inhalation of infective spores found in areas laden with decaying bird or bat droppings, such as well established roost sites. Symptoms of histoplasmosis include fatigue, fever, and chest pains. Avoiding these conditions is the best way to avoid infection.

Cryptococcosis (krip-toe-coc-o-sis) is a fungal disease caused by *Cryptococcus neoformans*, which can be found in the droppings of birds, or *Cryptococcus gattii*, which grows in the soil around trees. Persons who are immunocompromised are much more likely to become ill from exposure to these fungi than those with a healthy immune systems.

Ringworm is a fungal infection characterized by a raised, ring-shaped, red rash. It is transmitted to humans from contact with spores, or the infected hairs or dermal scales of infected animals.

Avian chlamydiosis, often referred to as psittacosis, is different from the human-to-human pathogen. It is commonly found in pigeons, ducks, raptors, and psittacines (parrots). The organism is present in the tissues, nasal discharges, and droppings of infected birds. It is typically transmitted to humans through inhalation of fecal dust. Clinical signs include fever, chills, and respiratory difficulties.

Viruses

Contageous ecthyma, also known as "orf," is a viral infection caused by a member of the poxvirus group. It occurs mostly in sheep and goats but can infect wild populations of artiodactyls. It has been found in bighorn sheep, Dall's sheep, Rocky Mountain goat, and caribou. The infection usually begins with the formation of blisters that become crusty scabs – typically on the lips and muzzle. As the sore heals, the scab falls off. The virus contained in the scab can remain infectious for years in the right environment. Transmission to humans is usually through contact of abraded skin with the sores or infected material. Non-porous gloves should be worn when handling potentially infected animals.

Hantavirus can be transmitted through inhalation of the aerosolized saliva, urine, or feces of infected rodents. Rarely, the virus can also be spread through ingestion, bites, and broken skin. Symptoms include flu-like conditions and respiratory distress that can lead to what is called hantavirus cardiopulmonary syndrome.

Mosquito-borne encephalitis refers to viral diseases transmitted by mosquitoes, including West Nile virus, St Louis encephalitis, and western equine encephalomyelitis. Each of these viruses is maintained in nature through a mosquito – bird – mosquito life cycle. Symptoms in humans include fever, chills, and myalgia.

Rabies is a deadly disease caused by a neurotropic virus. There are various strains of the virus that are maintained in particular species, or reservoir hosts, and usually confined to a geographic region. In the United States there are variants for raccoon, bats, skunk, fox, and coyote.

Typically, the rabies virus is transmitted through a bite or where saliva or body fluids from an infected animal come into contact with open skin or mucous membranes. Although post-exposure treatment can be effective, once the disease is contracted it is nearly always fatal. Protective equipment and proper restraint methods can reduce the chance of being bitten. Rabies pre-exposure vaccinations should be a requirement for anyone routinely exposed to rabies vector species (RVS).

Parasites

Larva migrans is a condition in which larval stages of a parasite, such as a roundworm, persist and migrate through an animal's body. The raccoon roundworm, *Baylisascaris procyonis*, and the canine and feline roundworms, *Toxocara* spp., are well known causes of the condition in humans.

B. procyonis lives in the intestinal tract of the raccoon. A single adult worm can produce close to 1,000,000 eggs per day and an infected raccoon can shed 45,000,000 eggs daily through its feces. The eggs are protected by a hard cuticle, making them nearly indestructible – they can survive for years, even in a harsh environment.

After an egg is shed in the feces, it takes between 2 and 4 weeks for the larvae to develop, when the egg will be infective. When an embryonated egg is ingested by another animal it hatches and the larva begins its journey, migrating erratically through the organs, eyes, and central nervous system. In small animals, such as rodents or birds, the infection can cause central nervous system disorders, illness, and death. When a raccoon consumes one of the infective eggs or an infected carcass, the larva completes its life cycle, returning to the small intestine to become a mature roundworm (Fig. 1).

Diagnosis in humans can be extremely difficult and there are few treatments available. Prevention is the key. Hygiene, protective clothing, gloves, and strict policies for decontamination will reduce the chance of infection.

Echinococcosis, or hydatidosis, is another enteric infectious disease. It is transmitted through ingestion of infective tapeworm eggs shed in the feces of an infected animal, particular a wild canid. It can cause parasitic tumors on the liver.

Giardia lamblia is an infectious agent found in beaver, muskrat, and waterfowl. Transmission is through ingestion of the cysts acquired from infected water or from the feces of an infected animal. Symptoms include chronic diarrhea, bloating, and frequent loose, pale stools.

Toxoplasmosis is caused by ingestion of the oocyst of the protozoan *Toxoplasma gondii*. It can occur through ingestion of undercooked meat containing the cysts, or from ingesting the toxoplasma eggs shed in the feces of the common housecat. Cats become infected through contact with wild species. While other animals can become infected, only the housecat sheds the eggs. The disease is seldom severe in people with healthy immune systems, but it can be a serious issue for pregnant woman. Each year in the United States, up to 3000 babies are born with ocular lesions caused by *T. gondii*. Preventing cats from exposure to wild species will reduce the potential for infection.

Sarcoptic mange is the highly contagious parasitic skin disease of dogs caused by the dog-specific mite called *Sarcoptes scabiei* var. *canis*. In humans, *S. scabiei* var. *canis* can cause skin irritation that usually heals within three weeks. This mite infestation is different than demodectic mange in dogs, which is caused by proliferation of a hair follicle mite that is part of a dog's normal skin fauna, kept in check in healthy dogs. Dogs with weakened immune systems can become overloaded with the parasites.

After handling birds or their nesting material, you may find that you have bird lice crawling on your skin and in your hair. The largest and most alarming, due to their size, are pelican lice. The creatures pose no real threat to humans and die fairly quickly. A good shower with soap and water and a change of clothes should be enough to rid yourself of these arthropods.

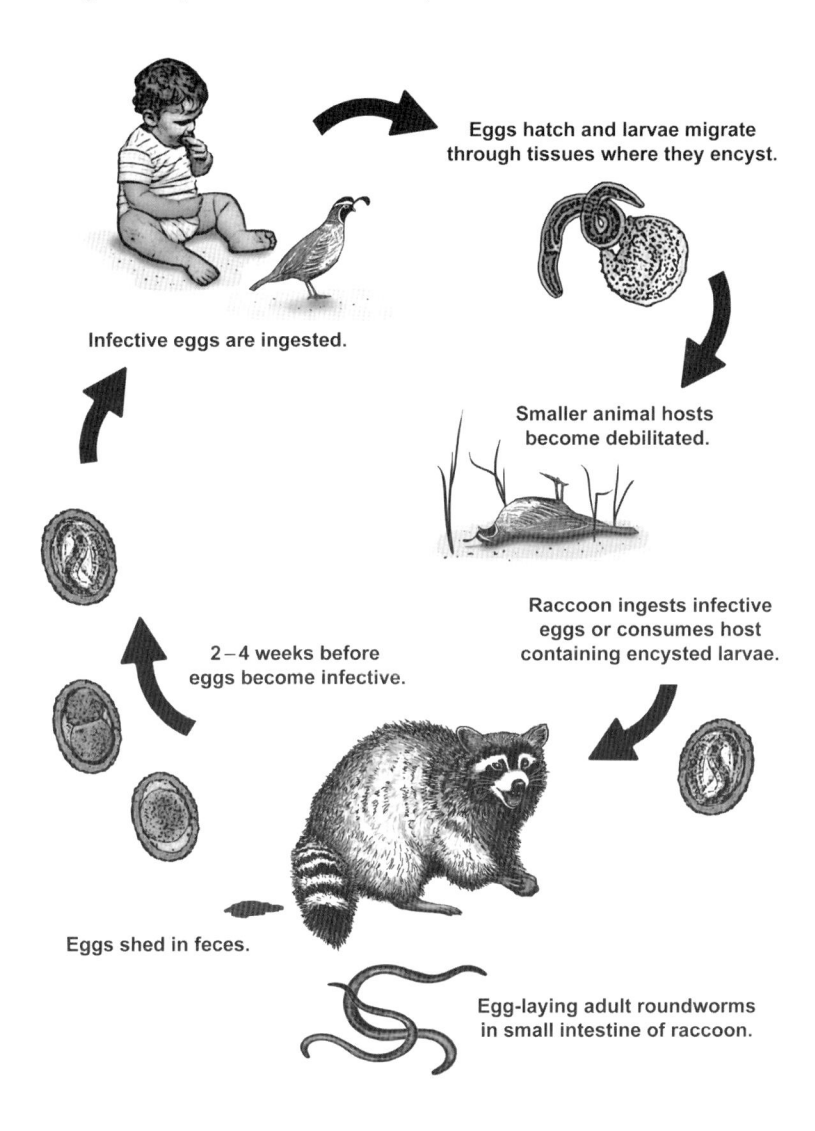

Fig. 1 The life cycle of *Baylisascaris procyonis*, the raccoon roundworm.

Personal protective equipment

In the course of duty, responders will encounter a variety of potentially hazardous conditions requiring specific personal protective equipment (PPE). The selection of safety gear is generally based on the task, environmental conditions, and the species of animal involved.

Choosing the right equipment is critical. Supervisors will want to oversee proper selection of gear, and make sure personnel receive adequate instruction and training to ensure PPE is effectively used.

Harmful substances enter the body through ingestion, inhalation, absorption, and injection – through punctures or bites. Barrier protection is a principle means of protecting workers from the risk of injury or infection.

Gloves are imperative when working with animals or with contaminated material. They are one of the most important articles of safety gear for a wildlife rescuer. Gloves must be thick enough to adequately protect the wearer but not so thick that they compromise use of equipment or safe handling of an animal.

Exam gloves, usually made of latex or rubber, are thin and offer the least amount of protection from punctures and bites. They also offer little protection from chemicals. Nitrile gloves, which are made from a synthetic polymer, are slightly thicker and more durable, and offer handlers protection from certain chemicals. When wearing these thin safety gloves, it is wise to wear two pairs, especially if working with hazardous material – because you can remove the outer layer and still be protected.

Leather gloves are porous and will not offer much protection from chemicals or fluids, but they do offer good protection from scrapes and punctures. Wearing exam gloves underneath is recommended.

Leather gloves come in a variety of thicknesses and lengths. Gauntlet-style gloves, like those made for pruning roses, offer heavy cover for the arms, with softer, more pliable leather protecting the hands. General workers' gloves and welders' gloves are made of thicker leather, providing a good barrier against punctures but sacrificing flexibility and dexterity. Commercially produced animal handling gloves can be costly but offer superior protection and versatility.

A worker's eyes must also be protected and, again, one must take care to select appropriate protection that will not compromise vision. Safety glasses come in a variety of shapes, sizes, and colors. Specialty safety glasses can be found with bifocal magnifiers or made to fit over prescription eyewear. In some cases, eyewear will not offer enough protection; a face shield may be necessary to protect workers from harmful splatters or sprays or from long-necked birds that intentionally strike at the face.

Facemasks can be used to shield the mouth and mucous membranes of the nose from potentially harmful droplets. The lightweight, three-ply non-woven medical or surgical masks are known to hold up well, better than paper masks. However, these masks will not guard against fine particulates or airborne pathogens. When there is a potential for aerosolized dangers, consider wearing sufficient respiratory protection, such as an air-purifying respirator. Persons with health issues, though, may not be fit to use a respirator and should consult with their physician first.

Filtering facepiece respirators are worn over the nose and mouth, secured by elastic headbands. They come in a variety of shapes and sizes. Some are designed for single or short-term use, such as the disposable N95 (US Standard) particulate respirator. The foldable styles of these single-use facemasks are recommended over the pre-molded variety for versatility and fit.

Half-mask air-purifying respirators are made of flexible rubber or plastic and have replaceable cartridges. They are held in place with adjustable headbands. The cartridge filters must be checked and replaced as needed.

With regard to footwear, it is extremely important that it be specific to the activity and fine-tuned for comfort. Rescuers may be on their feet for many, many hours. Whether walking a flat sandy beach, traversing a rocky trail, or managing an urban jungle, the footwear must be suited for the terrain, providing the wearer with stability, support, adequate traction, and protection from the elements. Footwear must also be properly fitted to the user. Properly fitted shoes will feel comfortable and less apt to cause issues such as blisters. Even so, newly purchased footwear, like hiking boots, will need to be "broken in" before it is worn on a long rescue mission.

In some environments, spatterdashes, or gaiters, may be advisable. These are leg coverings that help protect the foot and lower leg from branches, thorns, mud, and snow. They also help prevent debris, ticks, and chiggers from entering through the top of the shoe. They are made from a variety of materials and can be found in various lengths.

When working under conditions where there is a risk of being struck in the head or falling, consider wearing a protective helmet. Safety helmets come in a wide variety of shapes, sizes, weights, and thicknesses. Again, the choice of helmet will depend on the potential danger.

When working from a boat or around deep or swift moving water, a personal flotation device (PFD) should be worn. It is important, though, to select the appropriate type.

Every life vest or life jacket will have a Coast Guard rating for its life-saving potential under stated conditions. When selecting a PFD, the rating should be appropriate for the wearer's size and weight, and the type of water they will encounter. A Type I PFD offshore life jacket offers maximum buoyancy, enough to turn an unconscious person upright in the water. These models are usually bright colored, bulky and cumbersome. Type II PFDs, sometimes referred to as "horse collars," are near-shore life jackets intended for use in inland waters. They easily slip over the head and around the neck. Type III PFDs are much more versatile and more comfortable to wear. These flotation aids can be found in a variety of shapes, sizes, and colors and are tailored for a variety of outdoor activities. A Type IV flotation device is a throwable device, such as a ring or buoy, meant to be held onto rather than worn. Type V PFDs are called "special use" or "hybrids," meaning they are intended for specific activities.

Modern inflatable PFDs received their US Coast Guard rating in the 1990s and are gaining in popularity. They are lightweight, comfortable, and suitable for many activities. They come in Type I, Type II, and Type III PFD specifications. One of their drawbacks is that they require regular user checks and maintenance.

Additional protective gear will be required when working in or around potentially hazardous material, such as petroleum. As with all other safety gear, workers must select the appropriate level of PPE, depending on the working conditions, their duties, and the hazardous material they will be exposed to. Assigning levels of PPE is beyond the scope of this book; however, the following is an overview of PPE commonly issued during oil spill response in the United States.

Protection from hazardous materials

In the United States, the agency known as the Occupational Safety and Health Administration (OSHA) is responsible for setting and enforcing standards related to workers and their work environment. It administers the Occupational Safety and Health Act of 1970, which requires employers to be responsible for safe work environments for their employees. When wildlife rescuers are involved in recovering contaminated animals, they will be required to follow a set of guidelines. These safety requirements might be imposed by the organization they're representing or by the lead agency taking responsibility for response and clean-up efforts.

Most wildlife rescuers will not find themselves working in what is referred to as the "hot zone," where exposure to extremely dangerous conditions is high. Following oil spills, rescue operations are typically conducted along nearby stretches of shoreline where contaminated animals will seek shelter from clean-up crews. While searching for animals, rescuers might find patches of viscous oil, or washed up globs that look like black rubber pancakes flecked with sand. Once petroleum is exposed to the elements – air, water, sunlight – it starts to degrade. Even so, weathered oil can still be hazardous.

Crude oil may consist of hundreds of different compounds. The subset of "worse-case" chemicals include the volatile compounds benzene, toluene, ethylbenzene, and xylenes, referred to as BTEX, and the polycyclic aromatic hydrocarbons, or PAHs. Several of these compounds, including benzene, are known carcinogens.

When wildlife rescuers are at risk of exposure to oil by contact with oiled animals or moderately oiled shorelines, they will probably be required to wear what is called modified Level D protective equipment. This can include coveralls made of a material called Tyvek.

Tyvek is a lightweight synthetic material made of high-density polyethylene. One of its unique qualities is its ability to allow water vapor to pass through it while preventing liquid water from penetrating. Tyvek suits are normally white.

Additional protective gear typical for Level D workers can include protective steel-toed rubber boots, two sets of chemical resistant gloves, such as Nitrile gloves, and eye protection. Depending on the work environment, they may also be required to wear a hard hat or a life jacket.

To create a complete protective barrier, the sleeves of the Tyvek coveralls are taped to the outermost set of gloves (Fig. 2). Similarly, each pant leg of the suit is taped against the outside of the worker's boots. Leaving a tab on the duct tape will help when it becomes time to remove the suit.

When working on an oil spill, personnel should be prepared to decontaminate. In the field, this process usually involves a systematic removal of the soiled Tyvek suit while standing on a tarp or sheet that can be rolled up and stowed. The last article of protective wear to be removed should be the innermost set of

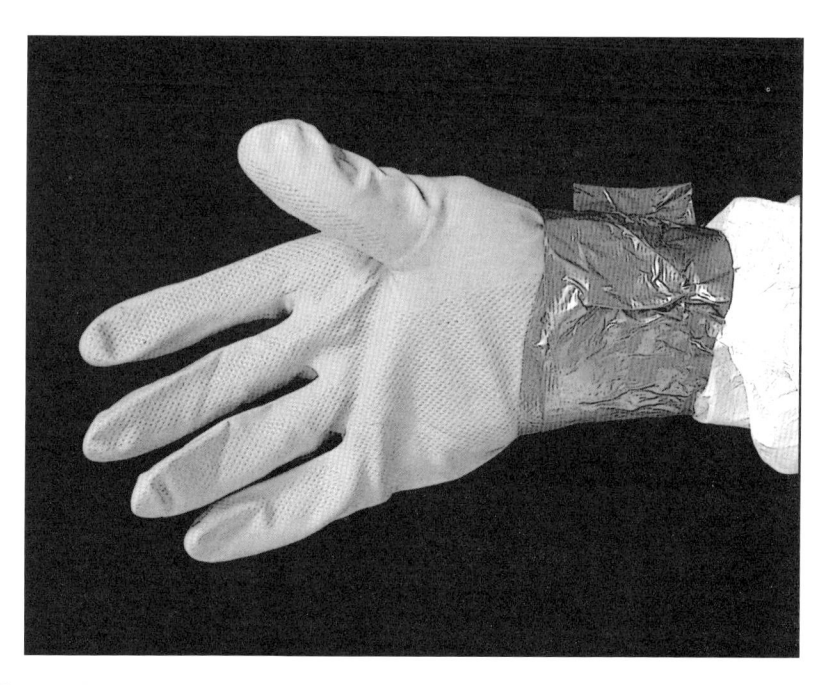

Fig. 2 The Tyvek suit is taped onto the outside of gloves and boots to protect against exposure to hazardous material.

gloves, rolling them off and onto themselves. If boots are soiled they must be decontaminated or, at the very least, they should be isolated to eliminate cross contamination. A mixture of rubbing alcohol and tea tree oil, or melaleuca oil, has been used in the field to successfully remove oil from hands and rescue gear.

Basic safety and preparedness guidelines

Every rescue mission will present a unique scenario with different hazards and varying degrees of risk. Observing a few basic principles can greatly enhance overall safety, no matter what the situation.

One of the first basic rules is for rescuers who are working in potentially hazardous situations to work in teams of two or more. This is called the "buddy system." The partners assist and protect each other, working together to avoid or mitigate potentially dangerous situations. Where they may become separated – out of sight of one another – they must be equipped with a reliable means of communication.

When working in the field, rescue teams must be equipped with suitable functioning communications equipment. They must be able to communicate back and forth with each other and with the "outside world" – it is imperative they have a reliable means of calling for assistance or making contact with their headquarters. As a safety precaution, rescuers may be required to check in with an off-site supervisor before beginning an assignment, and again once finished and they have left the scene. Communication equipment must be suited to the environment. This might mean that workers are equipped with close-range two-way radios and cell phones, or in very remote locations a team may require VHF radios and satellite phones.

Another essential for safety is the practice of cleanliness when working with or around animals. It helps reduce the spread of contaminants and disease. Rescue equipment should be thoroughly cleaned and disinfected after each use.

To help them stay clean, rescuers workers should assemble a kit that contains a change of clothing, plenty of exam gloves, a disposable mask or two, a few large heavy-duty garbage bags, rags, tissues, alcohol-based antiseptic gel, and a container of anti-microbial wipes. Wipes are preferred over gels as they help to mechanically clean the skin. Hand washing, though, is preferred over hand rubs or wipes.

Hand hygiene is one of the most important measures in controlling the risk of disease transmission. The transmission of many zoonotic diseases can be prevented simply by hand washing with soap and water – even rinsing them in clear water, if nothing else. Anyone working with or around animals must get into the habit of washing their hands thoroughly between and after contact with equipment or potentially contaminated material. After washing, rescuers should dry their hands with a disposable towel and use the towel to turn off the faucet. Hand jewelry can collect dirt, reducing the effectiveness of washing, so jewelry should be removed beforehand. Additionally, workers' fingernails should be scrubbed clean and kept short.

Ideally, the above precautionary measures should be part of a larger set of safety guidelines. Additional safety policies should include procedures for handling soiled laundry, animal quarantine guidelines, proper use of disinfectants, handling of used needles, procedures for handling hazardous material, evacuation plans, and vector control policies.

In addition, wildlife rescuers should become familiar with a very practical system for identifying and controlling risk. It is called "operational risk management" (ORM). While ORM is commonly associated with rescue missions, this valuable tool can be integrated into all organizational levels of a wildlife rescue program, enhancing effectiveness and reducing accidents.

Operational risk management

The key to safety is awareness and preparation. ORM is a formal, methodical, cyclic process used to identify hazards and mitigate risk. Mindful discovery of risks results in a plan of action that anticipates problems that might arise and predetermines methods of dealing with them. This pre-emptive, rather than reactive, approach to managing risk is used in the planning stage and is to be applied continually throughout a rescue mission, as resources, the environment, even the mission itself, can change.

The ORM process is based on this set of fundamental principles: accept no unnecessary risk; accept risk only when the sum of benefits exceeds the sum of potential costs; risk decisions must be made at the appropriate decision-making level; and anticipate and manage risk through planning.

While in-depth planning for a rescue mission is optimal, during emergency situations there is usually little time to spare. Even so, a time-critical application of the ORM process will aid rescue personnel in making sound decisions "on the fly" through a verbal or mental review of the circumstances. As time permits, rescue personnel will want to exercise a more deliberate review of circumstances through brainstorming and extensive planning, whenever possible. The ORM process can also be applied more strategically in long-term planning of complex missions, utilizing research, testing, and data analysis.

The process of assessing and managing risk can be tailored to the needs of an organization or agency. Ideally, however, the process will include the following steps: (1) define the mission or task; (2) identify the hazards; (3) assess the risks;

(4) identify control options; (5) re-evaluate risk versus gain; (6) execute decisions; (7) monitor and watch for changes.

The process begins with defining the mission and breaking down the operation into "bite-size" stages. For example, in the rescue of a coyote tangled in a barbed wire fence in a field bordering a well traveled highway, the goal would be the careful untangling of the animal and its transport to definitive care. The mission could be broken down into these steps: (a) muster rescue crew on site; (b) traverse field to animal, untangle it from the wire and place it into a carrier; (c) carry the animal crate across the field and load it into the awaiting transport vehicle.

Step 2 calls for identifying the hazards within each stage defined in step 1. This is where members of the rescue unit should come together and brainstorm as a group. Anyone can fail to identify a hazard, which is why it is so important for the team to work together in identifying them. This also helps ensure that everyone on the team is aware of the potential dangers.

As the team works together to identify existing hazards, they will also want to consider potential failures – what might go wrong. The question "What if?" can help tease out the possible threats. When developing their list of hazards, it is important that they also note the cause, as this helps to identify controls and safeguards.

Using the coyote rescue as an example, the environmental hazards would include the dangerous roadway, the uneven terrain rescuers must cross, the rusty barbed wire, and the terrified animal. On paper, it might look something like Fig. 3.

Step 3 involves assessing risk for each identified hazard. Risk is the potential for loss. It can be assessed through the SPE Model (Severity, Probability, Exposure), which assigns a risk factor using this formula: risk = severity × probability × exposure.

Severity represents the loss or consequences of a mishap. This can include injury, property or equipment damage, mission degradation, reduced morale, and adverse publicity. Severity can be assigned a rating, for example, of 1 to 5, with 1 = none or slight; 2 = minimal; 3 = significant; 4 = major; 5 = catastrophic.

Probability represents the likelihood that something will happen. It, too, can be rated from 1 to 5 as follows: 1 = impossible or remote under any conditions; 2 = unlikely under normal conditions; 3 = about 50% probability; 4 = greater than 50%; 5 = very likely to happen.

Exposure can represent the number of people involved, and the complexity and duration of the operation. Exposure can be rated from 1 to 4 as follows: 1 = none or below average; 2 = average; 3 = above average; 4 = great.

After assigning risk values to the specific hazards, they can be ranked from highest to lowest. The following is an example of how these sums can be classified to ascertain the best course of action. A rating of 80–100 might indicate a very

STEP 1

Mission objective: Rescue coyote caught in barbed wire fence

    1) Muster crew on scene
    2) Traverse field
    3) Remove animal from fence / confine into carrier
    4) Carry and load into transport vehicle

STEP 2

1) Muster crew on scene:

    - Very dangerous, busy highway, high speeds, no place to park.

    - A vehicle collision might occur.

2) Traverse field

    - Freshly plowed, uneven ground - potential trips/falls.

    - Could be shot at for trespassing!
        SAFEGUARD: Permission from landowner

3) Remove animal from fence / confine into carrier:

    - Coyote bites handler(s)
      Cause: improper handling/PPE
        SAFEGUARD: Exp. Handlers / HD gauntlet gloves

    - Rescuer receives eye injury from clipped barbed wire
      Cause: uncontrolled wire / no face/eye protection

4) Carry and load cage into transport vehicle:

    - Uneven ground / carrying live weight - slips/trips/falls

    - No safe parking.

Fig. 3 An example of the ORM process being used to identify hazards involved in a rescue of a coyote caught on a barbed wire fence (steps 1 and 2).

high risk with a recommendation to discontinue or stop; 60-79 could signify high risk, requiring immediate correction; 40-59 might pose substantial risk, needing correction; 20-39 might represent possible risk, where attention is recommended; 1-19 might indicate only slight risk, which could possibly be acceptable.

What is most important in this exercise is not the assigning of numerical values but the discussions that are generated within the team during the process. Involvement of team members ensures that they are aware of the risks and are part of the informed decision-making process as they move on to step 4.

Step 4 involves further brainstorming to identify risk control options and safeguards for all hazards requiring attention. A risk control must change the risk by impacting exposure, severity, or probability.

Protective equipment usually helps to mitigate severity; training, experience and attitude adjustments can reduce probability factors; and exposure can be reduced by lessening the number of people involved, including bystanders.

There are a few additional risk control options the team can consider. One option is to transfer all or part of the operation to another entity. Alternatively, the rescue could be delayed until conditions improve, or be cancelled altogether. Another option, specific to wildlife rescue missions, is to try to haze the animal, driving it from a perilous location in the hope that it relocates to a safer, more accessible place. Available resources and time criticality often influence the choice of control options.

In addressing the risks associated with the coyote rescue, the highway hazards could be mitigated by requesting assistance from law enforcement in creating a temporary buffer zone around the rescue vehicles. As for the risk of injury from the recoiling barbed wire, the severity could be all but eliminated through use of the appropriate personal protective equipment (Fig. 4).

Assuming the proposed safeguards and risk controls are in place, step 5 calls for re-evaluation of risk versus gain. This stage is a "reality check," where the objective of the mission is reconsidered. In doing this, the chief decision-maker will consider the cumulative risks and long-term consequences of his or her final decision – a decision that is often subjective.

STEP 3

RISK = SEVERITY (1-5) X PROBABILITY (1-5) X EXPOSURE (1-4)

1) Vehicle collision: 5 X 4 X 4 = 100 (high risk)
2) Possible face or eye injury: 5 X 3 X 2 = 30 (possible risk)
3) Slips/trips/falls in plowed field: 2 X 3 X 2 = 12 (slight risk)

STEP 4

RISK CONTROLS

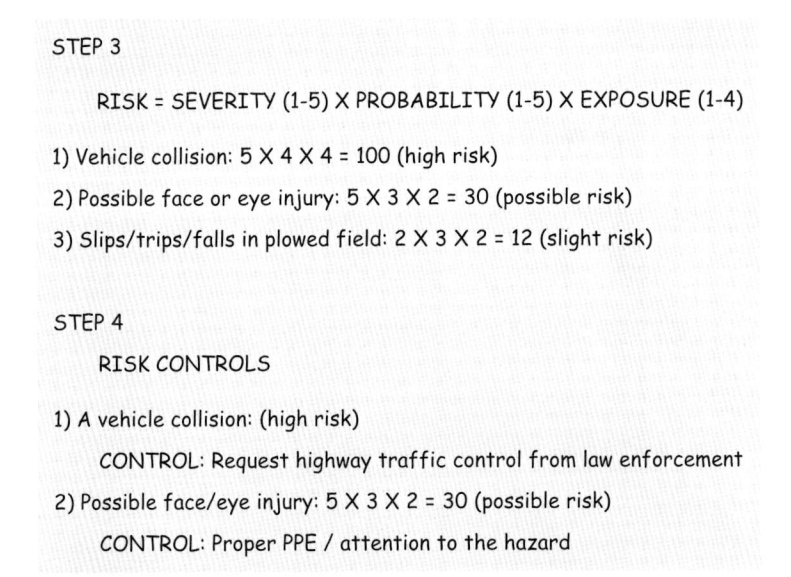

1) A vehicle collision: (high risk)
    CONTROL: Request highway traffic control from law enforcement
2) Possible face/eye injury: 5 X 3 X 2 = 30 (possible risk)
    CONTROL: Proper PPE / attention to the hazard

Fig. 4 An example of the ORM process being used to calculate the mathematical risk of injury and to identify risk controls (steps 3 and 4).

As a general rule, the acceptability of risk is based on a person's perception of risk, which is affected by the value they place on the anticipated loss. Decision-making associated with human rescues can appear simpler, more cut and dry, compared to non-human animal rescue missions where sentiments and value judgments can vary drastically. In the "animal rescue world," the decision to accept risk or spend resources to save an animal's life is often influenced by the decision-maker's personal feelings about the creature in peril. For example, a person might be willing to risk a great deal to save a raccoon but less interested in assisting a garden snake.

Step 6 is where personnel are informed of the risk management decision and the control measures are implemented. If there is objection from personnel, the decision-maker should be prepared to explain how they arrived at their decision, walking them through each step of the process.

The seventh, and final, step calls for monitoring of the situation, ensuring the controls are effective. The ORM is a continuous process (Fig. 5). As the mission is carried out and the situation evolves, rescue personnel will be watchful, returning to step 1 as needed.

Even though risk management concepts such as the one described above were developed for rescue operations, the process of identifying and evaluating risk

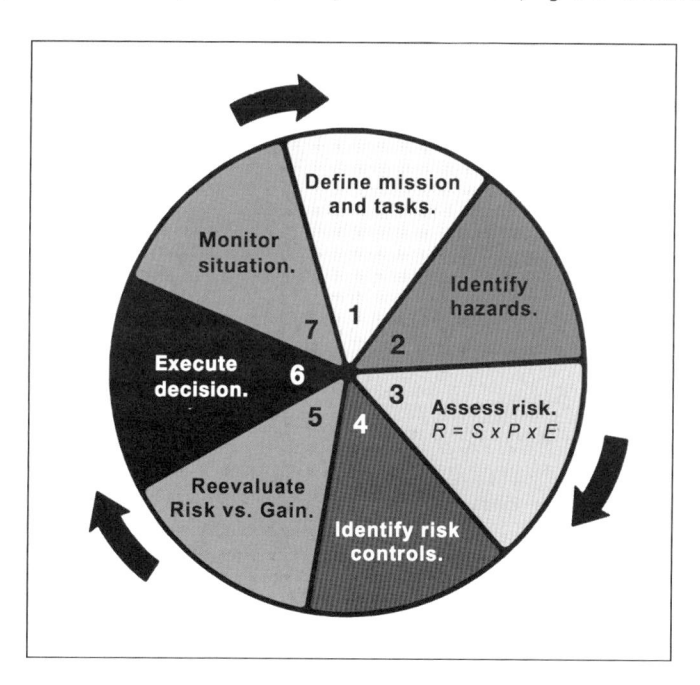

Fig. 5 The ORM is a continuous process.

can be integrated into daily activities, on and off duty, at all organizational levels. To simplify the process, workers can learn to use these five simple questions:

1 Why am I doing this?
2 What could possibly go wrong?
3 How could that affect others or me?
4 How likely is that to happen to me?
5 What can I do about it?

Outfitting

When called for service, rescue personnel should be prepared for their own personal safety and comfort needs. They may also be asked to provide their own gear, protective clothing, and improvisational tools for each mission. In preparation, workers should assemble their own individual "ready packs" or "go bags." These are backpacks, bags, or totes that contain essential gear and personal supplies specific to their assignments.

A 24-hour pack, for example, should contain supplies that will allow the rescuer to be self-reliant and comfortable for at least a day. It should be tailored to the environmental conditions they will be working under and the tasks or duties they expect to be assigned.

Standard contents of a 24-hour pack include alcohol-based gels and antimicrobial wipes, insect repellants, sunscreen, eye protection, gloves, a flashlight or headlamp, personal hygiene items, and a simple pastime, such as a book. A more detailed list of suggested contents is given in Appendix 1.

In addition to these supplies, rescue personnel should always carry plenty of water. Water can be carried in bottles, jugs, or bladder packs. Certain styles of backpacks have plastic bags, or bladders, that are refillable and allow easy access to water. Wildlife rescue workers should make sure to have an additional water supply available for personal hygiene, as cleanliness is one of the most effective ways of preventing the spread of contaminants and disease.

Since many of the diseases that can be transmitted from wild animals to humans can cause flu-like symptoms, rescuers should also carry a Medical Alert Card. The information on the card should include pre-existing medical conditions, drug allergies, and the species the rescuer has been exposed to. This will alert physicians to consider certain zoonotic diseases in their diagnosis. The information can be contained on a wallet-sized laminated card or, nowadays, the data can be stored on a flash drive.

Protective gear can be compiled in a separate container that a worker can take with them in their vehicle. This allows them to have an assortment of PPE to choose from. They will want to be sure to have safety glasses, a mask, and plenty of exam gloves.

Wildlife search and rescue workers will also want to have a few basic tools on hand, including a pair of scissors, wire cutters, pliers, and a knife. Duct tape, assorted cable ties, and a tarp are also considered standard. They can keep these in a separate bag or in a larger collection of handy items that is sometimes called a "hell box." A hell box is a container of assorted items that might come in handy for emergencies, repairs, or comfort. Additional contents might include string, heavy-duty clips, a compact sewing kit, safety pins, and super glue.

Depending on how often they are called for duty and the complexity of their missions, rescue workers might want to separate their packs for various uses. For example, they may want to have a small fanny pack with absolute essentials for quick, short-distance outings. They might have a medium-sized daypack prepared for longer journeys where they still have access to their vehicle, leaving heavy equipment and non-essentials behind. Wildlife paramedics will want to prepare a separate pack of personal items in addition to their medic's pack.

Wildlife search and rescue personnel should also prepare for the worst by having a full change of clothes on hand, including extra footwear.

In selecting the appropriate clothing for a rescue, one must consider the environment, terrain, weather, and level of activity expected. Since these conditions can change throughout a day, clothing must be versatile. A "layered system" allows rescuers to fine-tune for comfort. A five-layer system, often used for search and rescue missions, consists of undergarments, a wicking layer, a clothing layer, an insulating layer if necessary, and a shell to protect from wind, rain, or sun.

When choosing undergarments, plan for the unexpected and for extremes. For the ladies, dark-colored sport bras are recommended over light-colored, lacey nothings. With regard to socks, when trekking great distances two layers are essential – a thin sock against the skin, and a thicker, insulating sock on the outside.

When performing strenuous activity in cold climates, a wicking layer is recommended. This is worn over the undergarments. The purpose of this layer is to draw, or wick, perspiration away from the body and prevent the worker from becoming chilled.

For the clothing layer, apparel should be loose fitting or stretchy to allow for unrestricted movement. This layer can be versatile, with multiple closures to allow ventilation as needed. The fabric should dry quickly and be snag or tear resistant.

In warm to moderate climates, fatigues are ideal. They are lightweight, durable, roomy, and have multiple utility pockets. In cold climates, double-sided polar fleece bottoms and tops are form-fitting but stretchy, providing freedom of movement and protection from the elements. Fleece dries quickly, making it a good choice when working around water in a cold environment. Another option for cold weather bottoms would be a pair of insulated nylon cargo pants. Cotton pants are not recommended for work in cold weather.

Cotton, once wet, takes a long time to dry out. In cool environments this can be problematic, where silk and wool, on the other hand, offer breathability, wicking properties, and good insulation.

Synthetic fibers, such as polyester and nylon, come in a variety of weights for a range of activities and environmental conditions. Synthetic fabrics offer the rescuer the most suitable clothing with the greatest selection for optimizing a wardrobe.

When selecting colors, white or bright colors will stand out to teammates, onlookers, and wildlife. These might be the best option for human safety concerns. Darker earthy colors will be less conspicuous – a good choice when stealth is called for.

Shapes and patterns can also be used to break up the human form. A garment called a ghillie suit is an outer covering worn over clothing to help conceal the human form. They are usually made out of netting or open-weaved fabric with many small bits and strips of twine and cloth attached, often augmented with scraps of foliage from the area in which it is being used.

A rescuer's head may also need protection. In cold weather, a warm, insulating cap will help retain body heat. In warm sunny weather, rescuers may want to don a lightweight sports cap made of UV-treated nylon. Ideally, the cap's bill should have a dark underside to reduce glare. Side flaps and a cape will help protect ears and neck from sun exposure.

The welfare of the animal

The second component of wildlife rescue operations – and the second most critical thing to consider – is the safety of the animals. At the start of a mission, rescuers will want to determine what action will be in the animal's best interest. For example, if the animal in question is a healthy fledgling songbird that is just learning to fly, it might be best to leave it in the care of its wild parents. If, however, the animal needs to be captured and confined, rescuers must make every effort to minimize its pain and suffering in the process. This requires an understanding of how the wild animal perceives humans and the subsequent physiological effects this can have on its health and wellbeing.

Most wild animals perceive humans as predators. When a wild animal is captured and handled it does not comprehend the good intentions behind such mauling, only that its life is in jeopardy. As a result, the animal will respond as it would to a predator attack: its body gears up for the fight of its life. This innate response triggers a complex series of physiological changes that begin in the brain and may last until the animal is returned to health and liberty.

Understanding stress

Every living organism has an ideal internal state, a state of metabolic equilibrium that favors body function and survival. This is called homeostasis. The body is constantly working to maintain this optimum state (e.g. balancing the stimulating and tranquilizing chemicals in the body).

Sometimes, the body needs to react quickly – for example, to bolt from danger. For this purpose, animals have adapted a survival mechanism – the fight-or-flight response. This response acts like a switch, turning all vital systems onto "hyper" mode, when all physiological and behavioral functions are diverted towards the concerted effort to survive the threat. The fight-or-flight response is one component of the stress response and is the animal's first line of defense when it believes its life is threatened.

A stress response can be triggered by a perceived threat, a challenging condition, or a strong emotion – good or bad. Once the switch is tripped, the body responds quickly and automatically. For example, in an instant the heart will start racing and breathing will speed up.

The threat (real or perceived) that induces a stress response is called a stressor. Stressors can be physiological, psychological, or sociological, real or imagined. Quick, short-lived stressors are considered acute stressors, while chronic stressors are those that either persist over long periods of time or are a combination of many short-term acute stressors.

Stress is the psychophysiological state resulting from an animal's response to stressors. It is a state of being – a mechanism for coping. It is a subjective sensation, affecting individuals differently, with varied results depending on an organism's ability to adapt or cope.

For many, the word stress may conjure up negative emotional conditions such as frustration or anxiety, but stress is not always "bad." Exhilaration, joy, and enthusiasm are examples of stress-inducing emotions that have a positive influence. Acute stress is a survival strategy and this quick, short-term response is generally considered to be beneficial. Challenging exercise, mental stimulation, and competition are examples of stressful conditions that produce beneficial stress. It is when a stressor is severe or persistent that an animal may suffer distress.

Distress occurs when response to a stressful influence is so costly that it negatively affects the animal's wellbeing. Unable to adapt, unable to achieve and maintain psychophysiological homeostasis, the animal will suffer the deleterious affect of stress. Distress may come suddenly, through severe trauma, for example, or may develop over time as the animal is exposed to unrelenting stressful conditions.

The idea that stress could impact health was introduced in the 1930s by endocrinologist Hans Selye, when he noted elevated levels of hormones in his lab rats, related to the rodents' stressful environment. In studying how organisms react to stressful conditions, Selye formulated a model called the General Adaptation Syndrome, identifying three phases of coping. The first is the alarm response, in which the animal's body prepares for dealing with an imminent threat; the second is called the stage of resistance, where the animal's body tries to cope with a stressful condition; and the third is the stage of exhaustion, where its body's resources can become depleted. Most wild animals will go through a stress response when captured and confined, often over and over again, without relief.

In reaction to approach by a human, a wild animal may initially attempt to hide, or it may try to defend itself. This is a voluntary behavioral response. The animal's heart rate may increase and breathing will become accelerated as part of it experiencing apprehension. This may result in a stress load similar to a mild to heavy workout, depending on how much fear the animal experiences.

If the animal is startled or when it perceives the danger as a threat to its survival, the switch will trip, setting off a complex series of events within its body: the acute stress response. Within seconds, the nervous system responds by triggering the release of adrenaline (epinephrine) into the bloodstream. Adrenaline is the signal for the heart to beat more quickly and strongly, and for respiration to increase in order to get the body prepared for a quick escape. In addition, the brain sets off a cascade that results in the adrenal gland's secretion of the stress hormone glucocorticoid (in humans and most other mammals, this hormone is cortisol). The glucocorticoid stress hormones help the body to redirect all physiological functions toward handling and surviving the stressor, and it also prepares the body to handle the after-effects of the stress response. If the threat is eliminated, the body will return to its normal state, or homeostasis, within hours. If not, the animal's body mobilizes more of its resources to cope.

In certain circumstances, animals can habituate to a constant and consistent stressor. This can be a make-or-break phase. If the source of the stress is eliminated or if the animal is successful in habituating and thus achieving homeostasis despite the presence of what was once considered stressful, its body functions will eventually return to normal. If not, however, the effects of chronic stress will manifest and become problematic for the health of the animal.

This is the turning point from the "good" stress response to the "bad." While the acute stress response is an adaptive reaction that mobilizes energy and focuses body function to deal with an imminent threat, chronic stress occurs when an animal's body remains in hyper mode due to ongoing or recurring stressors, and its body starts to wear down. Not only are the body's resources depleted but prolonged exposure to the high levels of stress hormones can take a toll on the whole organism. For example, exposure to high levels of stress hormones affects the functioning of the immune system, and with its immune system compromised, a chronically stressed animal will be more prone to disease.

What this means for wildlife rescuers is that an animal's exposure to stressful conditions, repeatedly or extendedly, can be life-threatening. Even short-term handling can have serious consequences.

Research has established that wild animals can begin to lose weight immediately after capture. One such study, published in 1984 by Nicholas Davidson, showed that loss of body mass can occur in two phases. Within and up to 8 hours after capture, shorebirds lost body mass rapidly and steadily, despite the degree of handling. After 8 hours, the loss of mass continued but was more accelerated in birds that were handled.

The greatest proportionate weight loss can occur early on, within 30 minutes of capture, and may be associated with dehydration. What rescuers must keep in mind is that this physiological response can seriously compromise an already debilitated animal. By the time the animal reaches a definitive care facility, it can present with reduced body mass, severe dehydration, and altered blood chemistry exacerbated by the stress from its recent capture and confinement.

In a wildlife hospital or rehabilitation center, an animal can experience apprehension or fear each time it sees or hears its caregivers, experiencing the threat of imminent death every time it is handled for treatment – again, and again, and again. Even when a wild animal is seemingly at rest in its cage, the chances are it is experiencing stress overload from multiple stressors – foreign smells, alarming sounds, and the constant presence of predators. As a result, the animal's ability to heal will be impaired.

Perhaps one of the most important things for all rescuers and caregivers to understand is that, for the most part, the animals they will be dealing with will be in a state of extreme fatigue. Even though the animals may put up a good fight to elude capture, or be aggressive upon approach, they are exhausted – exhaustion, being a state characterized by reduced body function brought on by prolonged excessive activity and repeated exposure to powerful stressors. Carers must allow animals the time and space they require to regain their strength. The greatest supportive care they can provide might be privacy and quiet.

To limit an animal's stress is not to limit its care, but to be conscientious of influences that cause the animal stress, making changes wherever possible to eliminate or minimize those conditions. In a wildlife hospital setting, for example, if an animal must be fed, weighed, have its blood drawn and its cage cleaned, it would inflict less stress if these things were done all at once, leaving the animal to rest undisturbed for the remainder of the day.

If a wild animal remains in a state of panic while under care – if its environment is too stressful or it is handled too often, the resulting changes in its blood chemistry can completely undermine its convalescence. High levels of stress hormones in the bloodstream will reduce inflammation around wounds, prolonging the healing process and increasing the chance of infection. As the animal's muscles are metabolized to keep up with its body's heightened demands under stress, the animal will drop weight. Additional ill effects brought on by chronic stress include suppression of the immune system, elevated cholesterol, arteriosclerosis, gastric ulcers, alopecia, hyperglycemia, brain atrophy, and osteoporosis.

Faced with this dilemma of potentially causing additional harm to their charges, wildlife rescuers and caregivers must first recognize the levels of stress an animal may experience during capture, handling, and captivity so they may eliminate or minimize the stress-inducing conditions.

Minimizing stress during rescue operations

Inflicting some level of stress on a wild animal during rescue and rehabilitation is absolutely unavoidable. It is a part of the process that cannot be helped. However, steps can be taken to minimize an animal's suffering.

The degree of stress an animal experiences in the presence of humans will differ, between species and among individuals. Some young animals may have a higher tolerance for humans; prey species tend to be more nervous than predators; gregarious species may suffer added stress if housed alone; and dominant individuals tend to be more sensitive to captivity and handling than subordinates. Therefore, to take appropriate measures to reduce stressful conditions, rescue workers must understand the animal, and the species. They must be knowledgeable of its natural history – the species's unique physical and behavioral traits.

Some understanding of an animal's unique senses can be gained by observing its features. A wild hare, for example, with its large ears and bulbous eyes, will be highly attuned to visual stimuli and sound. Primary stressors would include loud noises, such as barking dogs, and movement. Its powerful back legs suggest short bursts of speed. Animals like this need to be able to bolt – it is hardwired into their design. When they are prevented from bolting, as when they are manually restrained, they experience tremendous distress. Such strain can cause something known as capture myopathy. It can also occur when an animal is chased down.

Capture myopathy, also called exertional myopathy, is a disease seen in animals during capture pursuit, trapping, restraint, and transport. It is the degeneration of muscle tissue caused by anaerobic glycolysis and the build-up of lactic acid brought on by extreme muscle exertion. High temperatures can exacerbate an animal's condition, speeding up the onset of capture myopathy. This can be caused by high ambient temperature or an animal's inability to cool itself after extreme exertion. Most at risk are animals that have been tranquilized after a chase.

The symptoms of capture myopathy, which include difficulty walking or flying, can appear within minutes or may take days to manifest. There is no cure for capture myopathy, only prevention.

To prevent capture myopathy rescuers should minimize pursuit time and avoid "chasing down" an animal. Physical restraint of wild animals should be limited, carried out using the appropriate number of experienced handlers, and with correct restraint techniques to perform the job efficiently and humanely, and only when absolutely necessary. In warm environments, capture attempts or restraint procedures should be planned for the coolest hours of the day.

As a rule, the amount of pressure used to restrain an animal should be no more than it takes to do the job. Additional touching will cause the animal greater stress. Audio and visual stressors should be minimized – an animal's eyes should be covered at all times. If muzzles or holding bags are used, handlers must make sure the animal is able to breathe normally. They must also keep watch for signs of hyperthermia – elevated body temperature resulting from handling and restraint. As an animal struggles during handling, its muscles generate heat, which can lead to an elevated body temperature, requiring immediate attention.

Animals can also exhaust themselves by struggling to get out of their cages. Newly captured animals will tend to fight incessantly to break free, striking the cage walls repeatedly. At least for short term or transport, the space they are confined to should be minimized to reduce the animal's momentum. However, rescuers must make sure there is adequate ventilation to prevent the animal from overheating.

Similarly, some animals will fight to get through wire caging or see-through containers. This behavior can be addressed by limiting their view of "freedom" with a sheet or a blind.

Some animals, such as rabbits or rails, will feel less fearful if they have something they can use as a hide. A draped towel, thick foliage, or a cardboard box inside the cage can help these particular animals to feel less exposed and more at ease.

To minimize stress, the substrate an animal is provided within their cage must not be slippery. A slick plastic or metal surface can be very distressful.

For most animals in captivity, the sights and sounds they experience will be tremendously frightening. The sound of a door closing, a telephone ringing, footsteps, the sight of a human in the distance – these are all stressors that can be minimized.

As a rule, wild animals should be housed away from human traffic. Blinds, tarps, sheets and towels, draped over or in front of cages, can reduce visual stressors. Loud noises should be eliminated, and human voices should be toned down and kept to a minimum around the animals. In some cases, sounds can be used to drown out human disturbances. For example, the hum of a fan motor or the trickling of a fountain can be used to mask human voices.

Another way for rescue workers to reduce stress is to respect an animal's natural photoperiod as much as possible, allowing the animal to rest undisturbed in a quiet and dark environment during the part of the day it would normally sleep.

When housing animals indoors, even temporarily, rescuers should also consider the type of artificial light the animals are being exposed to. Fluorescent light fixtures can emit a great deal of ultrasound. If they are older or have magnetic ballasts, as opposed to high-frequency electronic ballasts, they can produce a flickering that is perceptible by birds. In humans, 100 Hz fluorescent lighting has been linked to eyestrain, headaches and migraine, even though this rate of flicker is above the human perceptual flicker-fusion frequency.

Potential for success

The third component of wildlife search and rescue operations is an overall appraisal of the elements that go into making a rescue a success, or not. Success means no person was injured, no property was damaged, no animal was harmed, and the appropriate action was taken on behalf of the animal. This evaluation should focus on how prepared the responders are, considering current, as well as historical, preparedness. The elements can be grouped into the following categories: planning, practice, procurement, post-action, and proficiency.

Regarding planning, the success of a mission will be heavily influenced by both the amount and the quality of the planning that goes into it. Planning should involve all critical responders. Together they will confirm information, discuss the mission's objectives, identify hazards, develop risk management strategies, confirm how they will communicate throughout the mission, and draft a plan of action, which will include contingency plans for what might happen. The plan will also assign particular roles and responsibilities.

It is imperative that all response team members clearly understand what they are responsible for doing, as well as what they should not do. How well they understand their job and boundaries, and how effectively the team works together, will depend on the quality of the leadership. Rescue coordinators and team leaders must have excellent communication and leadership skills. They must also have the savvy to effectively manage response resources.

Planning for a rescue will entail some level of organizational design. Be it a two-person rescue team or thousands of emergency responders, a management hierarchy, or command framework, is essential. There must be one person who assumes ultimate accountability, with cascading management levels underneath, as needed. The Incident Command System (ICS) is a good example of a management system with clear lines of authority.

After a plan is drafted and roles are assigned, there may be a delay before the actual rescue takes place. During this time, things can change, including the weather, personnel, and available equipment. To reaffirm roles and responsibilities, clarify expectations, and reassert ground rules, team leaders should get into the habit of delivering a briefing right before a rescue operation begins.

The brief, as opposed to the debrief, is a briefing given prior to the commencement of a task or operation. It is a presentation of facts to inform others of the proposed plan. For those who may not have been involved at the planning stage, the briefing should include a review of what principles, policies, procedures, and lessons learned helped to shape the plan. This is also the time when the team leader or appointed safety officer will usually deliver a safety briefing.

When reviewing safety precautions, clarifying expectations, and reaffirming ground rules, team members should be reminded of "no-nos" – for example, if they are expected to "wait back until told." This is also the time for team leaders to identify what level of initiative is appropriate, if any, and from which individuals.

Practice is another influential element in a rescue mission. How much real world experience has each member of the team had performing similar missions? Have they had much hands-on practice with the equipment they're going to be using? These are important questions to ask when judging the team's level of preparedness for the mission at hand. In readying for a rescue operation, the team should do a "practice run" to rehearse roles and to get a feel for how the equipment will operate. It might be helpful to decide on certain pro-words, also referred to as procedural words. These are words or phrases that help keep communications short. For example, if the Team Leader called out "Pressure", it could signal the team to move in, applying pressure to drive the animal in a particular direction.

Every mission is an opportunity to learn and to teach, increasing the proficiency of the team. Rescue coordinators and team leaders should set the stage for this by fostering a positive and encouraging attitude, allowing less experienced team members the opportunity to put what they have been learning into practice. This is an important part of building expert teams.

Procurement refers to the assemblage of resources critical for the rescue – the appropriate equipment, adequate supplies, the right number of skilled personnel, and any necessary external resources. If these are lacking in any way, this will reduce the chance of a successful outcome.

Post-action refers to the debriefing process. Debriefs play an important role in overall operational performance, or success. A debrief is a formal "after-action" review and critique of the actions taken during a rescue mission by all personnel involved. For large-scale events, involving other entities and agencies, the after-action review is sometimes referred to as a "hot wash." The "lessons learned" help to improve future response operations.

Proficiency is a summary evaluation of the responders and their potential to complete the objectives of the mission successfully. From the top down, from the leadership to the external resources, do they have the necessary skills and experience? The following is a look at the expertise required for successful wildlife search and rescue operations.

The mindset of the hunter and the hunted

Human beings are evolved predators with excellent vision and keen observation skills attuned to hunting animals. These traits are what skillful wildlife rescuers depend on to be successful. When searching for an impaired animal, they are hunting. When pursuing the animal, they are the hunter. Accessing the mindset of the hunter aids rescuers in designing successful capture plans. Some say they are able to almost "feel" what an animal is experiencing, helping them in predicting its response to approach. Others describe a primal, almost feral, state of mind in which they are able to survey topography speedily and note an animal's weakness in a flash. The most successful wildlife rescuers combine this intrinsic savvy with the study of natural history.

The importance of natural history

A key component to a successful search and rescue mission is how well the rescuers know their quarry – how familiar they are with the animal they're attempting to locate and capture.

Natural history is the study of a species's unique physiological and behavioral traits. It is an animal's life history. Knowing an animal's natural history is key – a crucial instrument for sound decisions on the capture and care of wild animals.

Being familiar with a species enables rescuers to determine if an animal is behaving normally or not – if it needs to be rescued. Knowing an animal's particular strengths and weaknesses helps rescuers to take the right precautions for their safety and the animal's welfare. Being familiar with an animal's unique attributes – its temperament and heightened senses – helps rescuers plan for its capture, as well as what measures to take to minimize stress.

Even without actually seeing an individual animal, just knowing the species lets rescuers start planning a rescue. Search and rescue plans are often based on the time of day a species is active, what it eats, where it forages. Capture plans are also designed around an animal's predicted escape route – the direction it is likely to flee. Rescuers also take into consideration an animal's mode of escape. For example, a bird that can fly straight up in the air requires a different capture strategy than a bird that predictably dives, or one that must scuttle to become airborne.

Even over the phone rescuers use their knowledge of natural history to determine what type of animal is being reported and if it truly needs assistance. For example, upon hearing a caller describe a drowning seal waving for help, the rescuer would know it's more likely to be a sea lion jug-handling – resting at the surface of the water with its fins raised. Natural history also helps rescuers to determine the age of an animal. For example, the description of a crow-like bird with blue-gray eyes tells the rescuer it's probably a juvenile crow.

An animal's unique characteristics will also influence the choice of temporary confinement. For example, aquatic birds such as loons and grebes must be provided a cushioned substrate to avoid secondary injuries. Subtle modifications to the way an animal is handled or housed, based on its species-specific needs, can influence its chance of survival.

By knowing what is characteristically normal for the species and what is not, rescuers are able to locate and identify wild animals in peril. For example, a large raft of grebes offshore with greater spacing between individuals than is normal – many of them lifting out of the water to shake off their feathers more frequently than normal – can indicate they are having trouble with their waterproofing; possibly they are oiled.

These keen observational skills can become so honed that all it takes is a glimpse for rescuers to discern the species and its state of health. What they perceive in that instant, what they base their observation on, can be referred to as jizz.

Jizz is a word used to describe the essences of a living thing – the impression of an organism. The perception is based not on a single trait or combination thereof, such as color, shape, size, or movement, but on something more – something intangible – a discernable character, not characteristic – something definite but indefinable. The term is widely used in the birding world and gaining popularity among rescuers when they try to describe their observations. Ornithologist Thomas A. Coward was the first to introduce the term jizz in his book *Bird Haunts and Nature Memories* (1922), with an entire chapter devoted to the subject.

The fundamentals of the search

Search strategies will be built around species-specific behavior. Rescuers will want to consider an animal's normal foraging or hunting habits, setting up a capture plan based around the time of day and the location where the animal is likely to

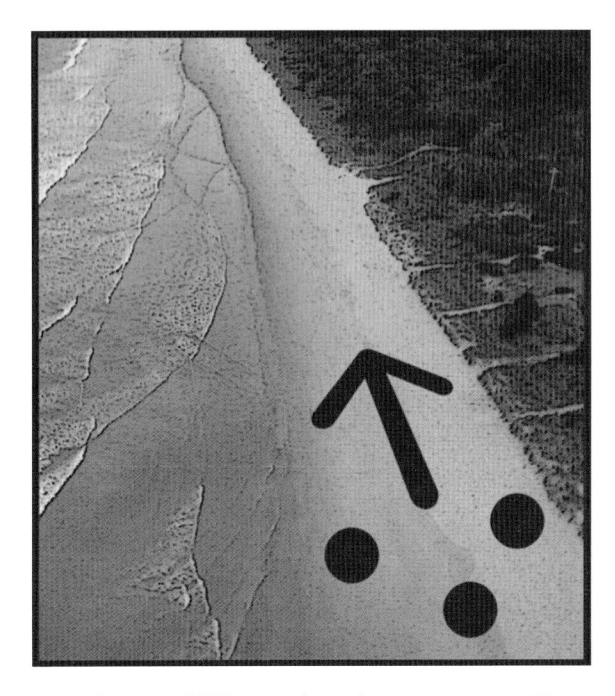

Fig. 6 The formation a wildlife search and rescue team will want to assume when canvassing a shoreline.

be found. Another strategy would be to look for the animal where it might go to rest, such as a particular haul out or loafing area. When animals repeatedly return to a specific location, this can be referred to as site fidelity.

When searching a vast stretch of shoreline for numerous animals – for example, after an oil spill – rescuers will want to canvas the breadth of the beach. To do so thoroughly, a team of three or more will want to line up in a crescent or "V" formation with the two "ends" leading (Fig. 6). As they move forward in unison, one searcher will cover the high ground while their teammates search the area in the middle and closest to the water, respectively. This formation offers effective coverage of an area and affords the capturers the greatest advantage should an animal flush from cover. This same "V" formation can be used in riverbeds and channels as well.

Searching does not always require a great deal of walking or a huge investment of time. Using high-powered binoculars or spotting scopes, searchers can cover vast stretches of coastline. In the process, they may identify "hot spots" where rescue efforts should focus.

For complex operations, such as oil spills, search teams may want to keep track of their field observations by filling out a Wildlife Observation Form like the one provided in Appendix 2. Here they can describe the terrain and access and egress points, with waypoints on important findings. This information can be extremely helpful in planning future coverage of those particular locations.

The fundamentals of the capture

Man has been chasing after and capturing wild animals for hundreds of thousands of years. The materials and equipment used to capture animals today may be contemporary, but the techniques and strategies for stalking or trapping animals are often simple and primitive, and based on an elementary set of rules.

Every capture strategy will be based on a few basic principles. One of the most influential factors will be the ability to predict the animal's direction of flight. Most wild animals want to avoid humans. Unless pressured to fight, most will react by moving away. The reaction will depend on the amount of "pressure" an animal perceives from the presence of a human. This varies among species and between individuals.

Rescuers must keep in mind that each individual animal possesses a unique character. Individuals will have a distinct temperament and peculiar habits. Among conspecifics, others of their own kind, many individuals will have an observable social ranking. These unique qualities must be considered when developing a capture plan.

Another thing for rescuers to be mindful of is that most prey species, and those that have recently evaded capture, will be highly sensitive to human presence. Animals that have been pursued over and over become nearly impossible to approach and continual harassment can force an animal to move out of an area altogether. Therefore, rescuers must take the time it takes to make the initial capture attempt a successful one.

In being able to predict an animal's flight *away* from a perceived threat, one is able to anticipate its travel *toward* safety. Some animals will flee to the nearest hide while others might rush to a body of water. For many species, this path of escape is so hardwired they will run right by a person with a net. As a rule, then, capturers should advance from the direction the animal will move toward. For example, when approaching a marine animal, capturers will want to stage themselves between the animal and the water (Fig. 7).

Another couple of basic principles apply to the final moments of a capture. Once within range for netting an animal or grabbing it by hand, the capturer should make his or her move first, before the animal does – this gives them the advantage of time and speed. It also catches the animal off-guard, instead of the other way around.

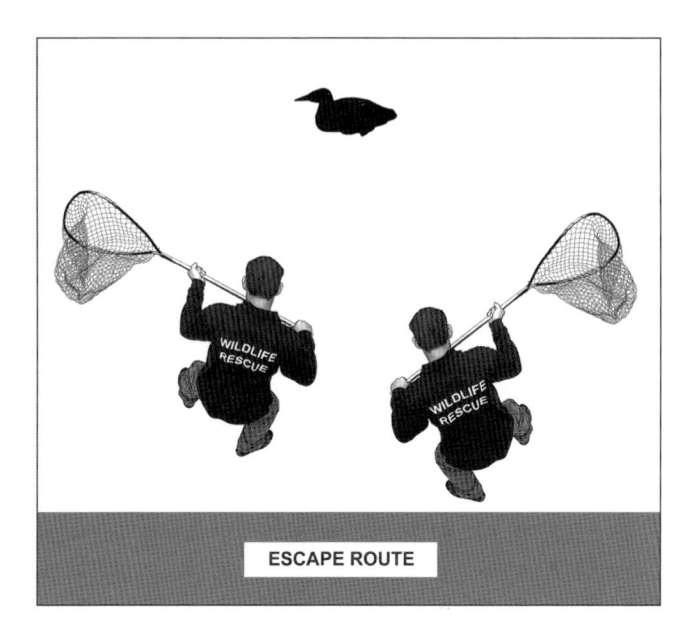

Fig. 7 As a rule, capturers should approach an animal from its route of escape.

The second rule is this: when the capturer makes the decision to go, they must give it their all and not falter. A split second of hesitation could cost them the capture.

Another important concept is that the capturer does not always have to have the animal in sight before making the final lunge, or sprint. Quite the contrary – if they can see the animal, the animal can see them. If the terrain allows a capturer to sneak up on an animal, they don't want to blunder the rescue by peeking at it from close range. Instead, they will want trust the animal is in the same spot, and go for it.

In these situations, where there is an opportunity to get close to an animal using features in the environment, it is helpful to have a second rescuer to direct the netter from afar, giving them feedback on changes in the animal's location or its behavior.

When the time comes for the netter to leap from cover, they will aim to intercept the animal as it tries to escape. Once the animal sees the netter, it will probably bolt. Instead of heading for where the animal was resting, the netter will sprint to where it will predictably be (Fig. 8). Rarely, but every once in a while, the element of surprise or intensity of the pressure is so great, the animal holds its ground and tries to defend itself rather than fleeing.

There are some instances where a hidden approach and a direct approach can be used together (Fig. 9). While this strategy is not suitable for animals with a

Potential capture zone

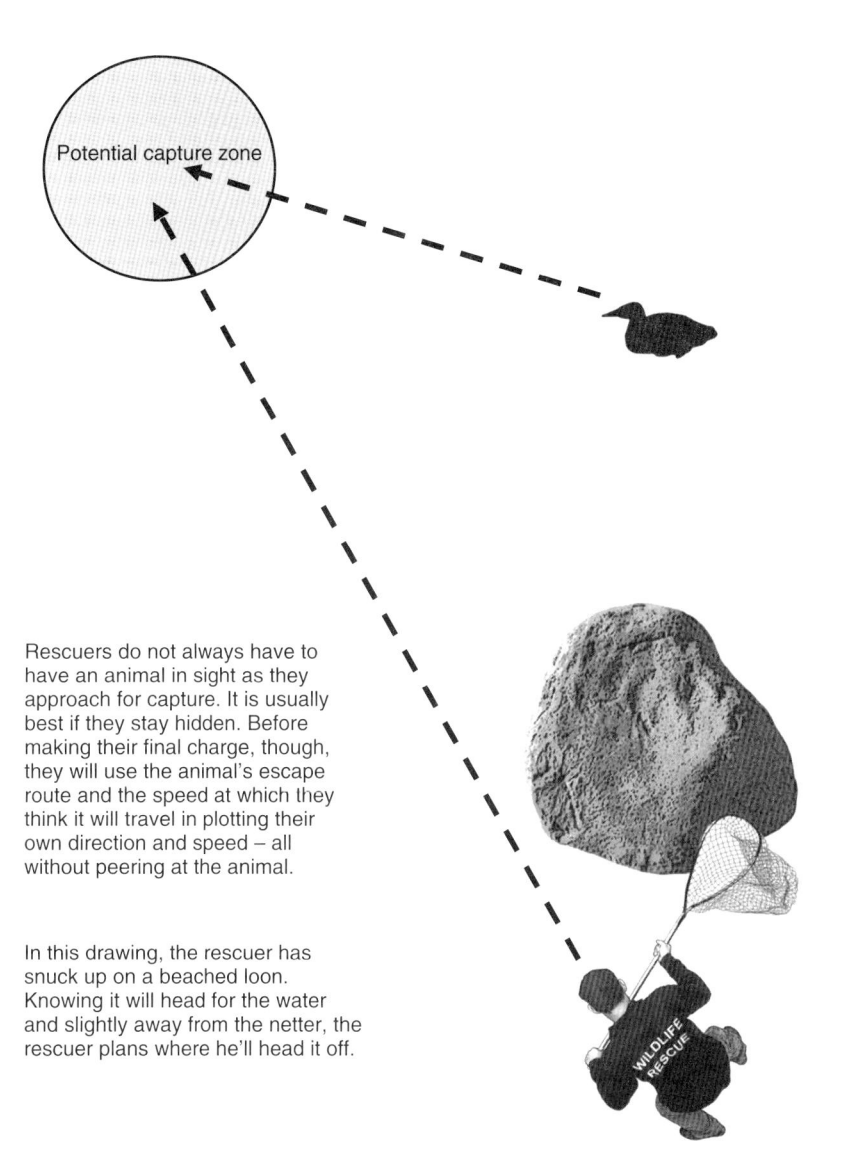

Rescuers do not always have to have an animal in sight as they approach for capture. It is usually best if they stay hidden. Before making their final charge, though, they will use the animal's escape route and the speed at which they think it will travel in plotting their own direction and speed – all without peering at the animal.

In this drawing, the rescuer has snuck up on a beached loon. Knowing it will head for the water and slightly away from the netter, the rescuer plans where he'll head it off.

Fig. 8 When capturers are ready to bolt from cover, they should plan to intercept the animal as it tries to escape.

strong drive to head in a specific direction – for example, marine animals – it can be a good strategy for those species that do not.

Another fundamental principle is that of pressure and how to use it to one's advantage. Some capture strategies will call for driving an animal in a particular direction. The animal's movement can be controlled by the amount of pressure

Fig. 9 In some situations it may be beneficial to have capturers approach from two directions.

that is applied. For example, just a glimpse of the rescuer may cause an animal to begin to walk in the opposing direction. By waving their arms, capturers will magnify the perceived threat, or pressure. One thing to note: two or more humans advancing from opposing directions will usually be more threatening to an animal than if they were to approach from the same direction.

Another standard rule is to use the landscape to one's advantage. Where there is an impenetrable boundary, such as a wall or a fence, this structure can be built into the capture plan, using it to block or redirect an animal's path of escape (Fig. 10).

Capture strategies that involve sneaking up on an animal require stealth and fortitude. Felines are excellent examples of how to approach stealthily – crouching, slinking, and making sudden advances only when their prey is distracted. To creep up on an animal, rescuers may need to skulk behind bushes, belly crawl, or freeze in a ballet-like pose for minutes at a time. Patience is key. It might take 40 minutes to inch close enough to capture an animal.

Fig. 10 Capturers can use existing barriers, such as walls or fences, to help them corner an animal.

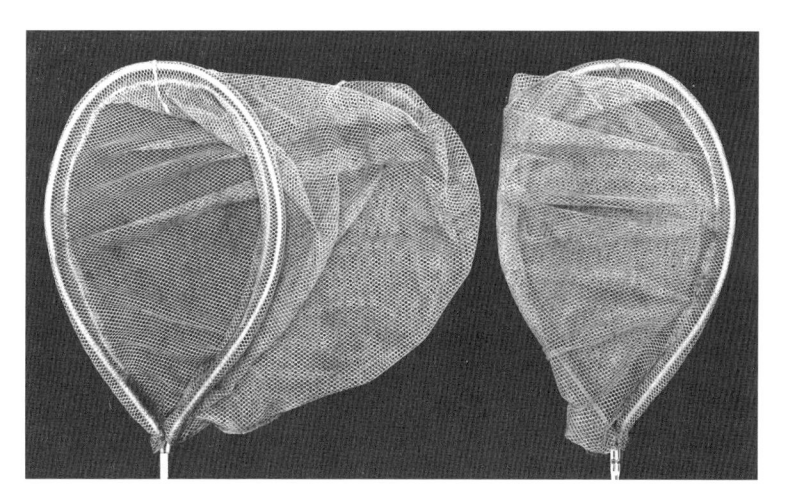

Fig. 11 Capturers must pay attention to subtle details that might scare off an animal they are trying to sneak up on. The bag of a hoop net can be folded on itself, preventing it from billowing in a breeze.

Fig. 12 An example of how capturers might conceal the hoop of their net as they approach an animal.

As they advance on an animal, capturers must pay sharp attention to details, because the animal surely is. They must be conscious of even the slightest movements. If it is windy, long hair should be secured so it doesn't flap in the wind. Bright, contrasting clothing, or billowy outerwear, especially material that makes noise, should be avoided. If the bag of their capture net is blowing about, it can be twisted so that the fabric is taut (Fig. 11).

When walking up on an animal, using a hand net, it is important for rescuers to keep a low profile – net down low and hidden as much as possible. The bag of the net can be concealed by the capturer's frame so that the animal only sees the person's figure (Fig. 12). This same technique applies when working in pairs.

Fig. 13 An illustration of how two capturers can reduce pressure on an animal by concealing their nets and the second capturer behind the lead capturer's frame.

The second netter will want to take a position behind the frame of the lead netter, keeping out of the animal's view as they advance in unison (Fig. 13).

In summary, each rescue will pose a unique set of circumstances to which rescuers must tailor their capture strategy, basing it will be based on the natural history of the species, the individual's condition and behavior, and some, if not all, of these a few these fundamental capture principles.

6 Anatomy of a response team

The following representation of a wildlife search and rescue operation assigns titles and respective duties to specific roles. While it accurately describes an existing response service, the titles are offered in example only.

A rescue begins when a person finds an animal they believe is suffering. It will be the telephone or the Internet they use to seek help. This juncture – this fleeting point in time when a person reaches out to help a wild animal, is paramount. This is where a wildlife rescue program has an opportunity to provide expert assistance.

This initial contact will be the rescuers' first opportunity to help the animal and the person who cared enough to call. Ideally an organization will have a Call Taker to receive calls and dispatch search and rescue personnel.

The Call Taker's first task will be to identify the caller's name and document both their callback number and complete address. This essential information is best recorded straight away. The Call Taker's second job will be to find out why the person called and what type of animal they have encountered.

Through a series of questions the Call Taker will tease out the facts. Most of the questions will center on the animal – what it looks like and how it is behaving. In this process, the Call Taker is performing an "initial assessment" of the animal. This evaluation is key in determining the best course of action.

A Call Taker's line of questioning might start with queries as to the type of animal, or what species it is. They will also ask questions about its condition. Is it bleeding? If so, from where and is it oozing or dripping? Is the animal standing, walking, running, or lying down? Does it appear conscious? Do its limbs appear symmetrical? Is it limping or otherwise favoring a limb? Is it obviously an infant or thought to be an adult animal?

To help with their evaluations, Call Takers should have the ability to receive digital images via the Internet. This can be extremely helpful in identifying the species and evaluating the health of the animal. Wider shots detailing the scene can be helpful in assessing the circumstances and for planning a rescue. Where possible, videoconferencing can be another useful tool.

Once the Call Taker has enough information for an initial assessment, he or she will address the animal's immediate needs, making sure to keep the caller out of harm's way. For instance, if the animal is cold, they might provide examples of how they can safely provide the animal with warmth. The simplest first aid measures can make the difference between an animal surviving and not.

After tending to the animal's critical needs, the Call Taker will focus on the broader picture, dispatching a team of specialists as needed.

Once the call has been transferred to field responders, the Call Taker's job is to provide them with backup, keeping them apprised of any changes that may take place before they arrive on scene. The Call Taker will also be relied on for additional support as the rescue progresses.

In the event that a search and rescue team is dispatched, the selection of personnel is usually made by the person overseeing the rescue efforts – the Response Coordinator. The Response Coordinator may be involved on scene, conducting the rescue mission as the Team Leader, or he or she may remain off-site, assigning a Team Leader to be in charge in the field.

Typically, search and rescue personnel converge at a predetermined location, designated by the Team Leader. Unless instructed otherwise, the crew will wait for the Team Leader to arrive before proceeding with the rescue operation.

Once the entire unit has converged on scene, the Team Leader will guide an assessment of the circumstances. If the animal's location is known, the team will assess its condition from afar, out of sight if possible. Wild animals, especially injured ones, behave differently in the presence of predators – humans.

When search and rescue operations are called for, it will be the Team Leader who will lead the rest of the responders in developing a plan. A good Team Leader will be charismatic, with just the right amount of assertiveness and refined communication skills. They should also have excellent situational awareness and be good at anticipating problems, helping their team plan ahead for exigencies.

Once the rescue team has established a plan and been assigned its duties, the Team Leader or the appointed Safety Officer will deliver a Safety Briefing.

After an animal is captured, if it requires immediate medical attention, the team's Wildlife Paramedic will see to its needs. Schooled in basic wildlife first aid and trauma care, Wildlife Paramedics play a major role in how many wildlife casualties make it from the field to definitive care. These medics will carry their own gear, separate from the other rescue gear and specific to the needs of the species they expect to encounter. A sample of a basic wildlife paramedic trauma bag can be found in Appendix 3.

The task of transporting a stable wild animal to a definitive care facility may be assigned to a special team – the Transport Team, trained and prepared to safely transport wild animals. The rescue mission should not be considered completed until the transporters have delivered the animal safely to a designated facility. Transporters should carry with them the proper paperwork, allowing them to legally transport wildlife. They should also consider some type of cage or vehicle placard, warning of their live wild animal cargo in case they are involved in a collision.

Sizeable wildlife rescue programs might have many skilled rescue workers with varying levels of expertise – a crew of search specialists, a team of capture experts, a specialized team of animal medics, a logistics unit, and a separate team of transporters, whereas a smaller program might have a handful or rescuers assuming multiple roles.

7 Overview of wildlife capture equipment

The towel

The average towel will be a wildlife rescuer's most valuable and most often used tool. Towels can be used to herd animals, to place over animals in order to pick them up safely. Towels are used to wrap and restrain animals, to cover their eyes, to drape over their cages, to line the bottoms of carriers. Towels, towels, and more towels – a rescuer's Must Have. Bed sheets and blankets also come in handy.

Herding boards

Herding boards, also known as "pig boards," are wooden shields that can be made out of plywood. They should be approximately 2.5 feet (75 cm) wide by 3.5 feet (105 cm) tall, with two handles on one side. They may also be fashioned with small holes for grasping or for seeing through. Herding boards can also be made out of heavy-duty plastic.

As the name implies, these boards are used to herd animals or to block their passage. They can also be used to conceal a rescuer when approaching an animal (Fig. 14). The boards are also used as shields to help protect rescuers from being bitten when working with an animal in a confined space or in restraint procedures to "squeeze" an animal so it may be examined or treated.

It is extremely important that the person assigned to a herding board is able to use it properly. It demands both mental and physical strength – wild animals can be surprisingly powerful and their protests can be violent. A boarder should be capable of wielding the armor in any direction, swiftly. In some cases they may need to be strong enough to sprint while carrying the board.

To use a herding board on a larger animal, a boarder can use one knee to help brace against the animal's might, leaning into the board as necessary (Fig. 15). The angle is important, too. The board should always be angled slightly in, toward an animal, not out, as this might encourage it to try climbing up and over.

When using herding boards to surround and contain an animal, four or more boards can be used to block it in, forming a box, basically. Once the animal is surrounded, a transport kennel can be placed between two boards. This is only an

56 Chapter 7

Fig. 14 Herding boards can be used to conceal a capturer.

Fig. 15 Boards are used to herd or block an animal's movement, and as shields to protect rescuers.

option if the animal cannot jump or climb onto the carrier. As the animal explores its options, the boards can be used to gently encourage the animal into the crate.

Nets and netting

Most nets are made of synthetic fibers, commonly nylon or polyethylene. The cords used in a weave will be described in weight, with the term denier, or in a measurement, noting the distance between intersecting pieces. Netting is used in building barricades, traps, and handled nets.

The hoop net

Hoop nets, also called hand nets, are usually made of strong, lightweight metal that is shaped into a ring, or hoop, with netting attached to form a bag. Handles can be short or long, collapsible, telescoping, or absent altogether. Manufactured nets come in a wide variety of shapes and sizes with different types of mesh bags.

As with any tool, having the right one for the job is essential. If the frame is suitable but the mesh bag isn't, the fabric can be replaced with more appropriate netting.

The material selected for the bag must be suited to the type of animal it will contain. It must not inflict injury or cause pain to the animal. For this reason, knotted fabric is avoided. As for the size of mesh, in general, the openings must be small enough that the animal cannot slip its head through. The cordage should be tight so it won't fray or snag on claws or teeth. When netting birds, the weave must not be so large that wings and feathers become tangled.

Tiny, delicate birds, such as hummingbirds and swifts, can be captured safely with a net made of a soft, lightweight mesh that is 1/16 inch (1.6 mm) or smaller, such as mosquito netting. Similar nets are used for collecting insects and handling pond fish. Especially deep bags can be made from this material to safely capture and secure reptiles.

A bag made of 1/8 inch (3.15 mm) mesh is suitable for capturing small birds and mammals. A larger weave of 1/4 inch (6.3 mm) can be used to secure larger songbirds, raptors, waterfowl, and some medium-sized mammals, depending on the strength of the fabric.

The greater the thickness and weight of the cordage, the less likely the fabric is to rip or be chewed through. The weight also helps in subduing an animal. Heavier weave is used to contain larger land mammals, such as coyotes or raccoons. Nets used to secure pinnipeds will typically vary from 1/2 to 2 inch (12.5–5.0 mm) knotless weave.

The open-ended hoop net

Over the years, a special net emerged, designed to reduce contact with dangerous animals. The open-ended hoop net allows rescuers to contain an animal in the bag of the net and transfer it into a carrier or enclosure without actual hands-on contact. This type of net can be made by replacing the material on a heavy-duty frame with a custom made bag.

For very large animals, like sea lions, this type of specialized hoop net can be made out of industrial grade PVC. Using a PVC heater, or bender, the warmed PVC can be shaped into a hoop. One method of doing this is to mark the desired size and shape on a piece of plywood. Nails can be used to help form the PVC to that shape (Fig. 16). Another way of forming the hoop out of PVC is to glue and screw 45 degree angled pieces together.

The bag for this type of net should be much deeper than a normal net – more like a tube. The bag can be woven to the hoop using nylon rope; it can be affixed permanently, or made to be detachable. Metal rings are attached to the end of the bag. A rope is fed through the rings, and the bag cinched closed using a quick-release type of knot. Additional drawstrings can be added as needed. The idea is to be able to confine the animal in the bag and safely transfer it to a carrier without touching it (Fig. 17).

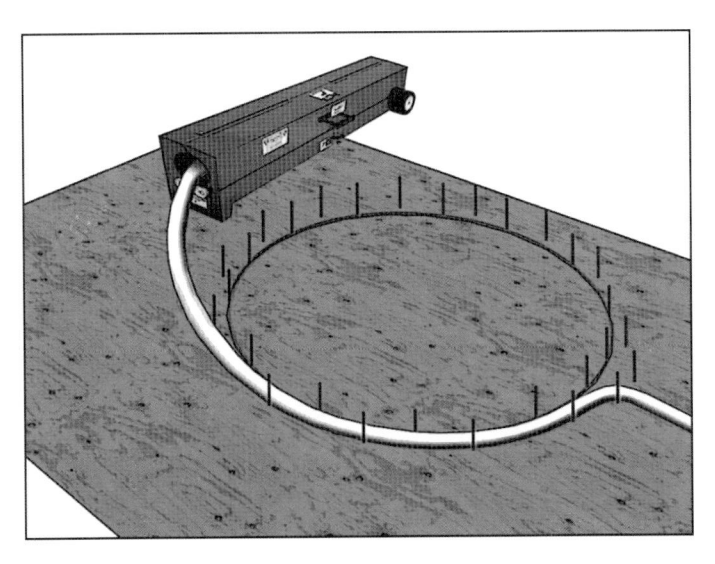

Fig. 16 An illustration of how to form the hoop of a net made out of PVC. The PVC is heated and shaped using a frame made out of nails hammered into a sheet of plywood.

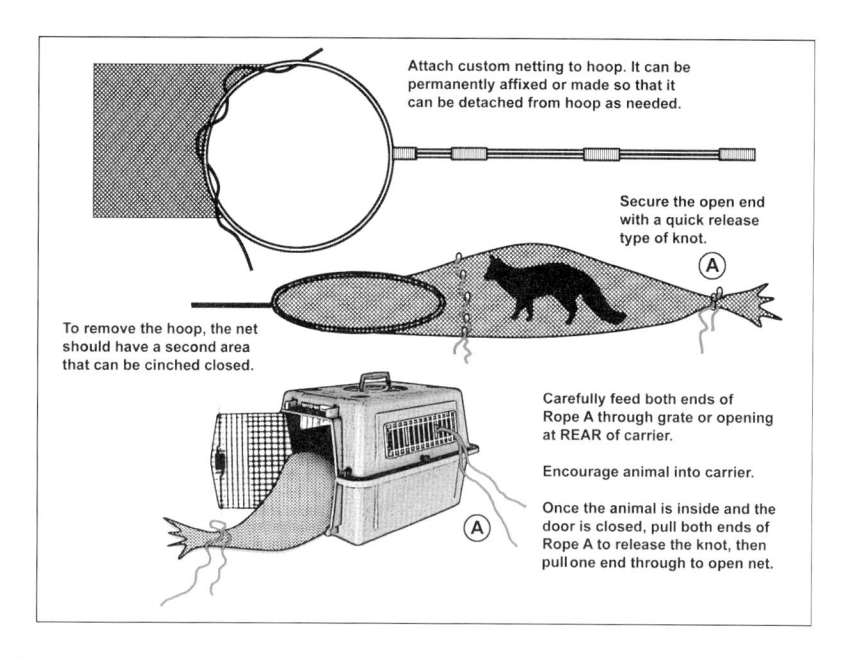

Fig. 17 Use of the open-ended hoop net.

When preparing to net an animal using this type of net, the netting material surrounding the hoop should be slack, not taut. This lessens the chance of it springing off an animal's body. The netter will aim to place the head of the animal in the center of the hoop, following through with a downward motion until the frame is flush with the ground. When necessary, the netter can gently and carefully drag the hoop a few feet (a meter or so), away from the animal's escape route, to encourage the animal to travel into the bag.

Typically, herding boards are used to surround the animal while it is in the mesh bag of the net, keeping it confined to the middle section until it is time to transfer it to a carrier. With enough boarders, the animal can be boxed in. Surrounding the animal with herding boards will help cut visual stressors, and its view of freedom. This can help reduce an animal's fight to escape.

To transfer an animal into a carrier, the long drawstrings at the end of the net are fed through an opening at the rear of the carrier. The tip of the net is drawn into the carrier as well, and held in place. Once this task is accomplished, the boarder stationed at that end of the net, closest to the carrier, will step aside. The animal is then encouraged through the tube and into the crate. When the animal is well inside, the door can be closed. The tip of the net can be dropped and the rope can then be drawn through the metal rings, releasing the animal inside the cage. The remainder of the netting can then be pulled out from under the door.

The throw net

A throw net is a circular piece of netting with a weighted rim, designed to be thrown through the air. When it is cast correctly, the net opens up as it spins toward the target animal. Capturers have experimented with creating rigid throwable rings using plastic tubing, PVC pipe, and "hula hoops," with mixed results.

Land seine

Netting is also used to create walls and barricades. A seine net is a vertically suspended net, like a curtain of netting. Used for fishing, a seine is suspended underwater, held up by floats and weighted at the bottom.

Land seines can be static – suspended like fencing – or triggered to spring upward. The choice of material depends on the type of animal being pursued. Lightweight garden netting can be used to capture small to medium-sized birds, or small mammals. This material is readily available at most hardware or garden supply stores and is relatively inexpensive. Stronger netting will be required for trapping heavier bodied animals.

The purpose of a static seine "wall" is to direct, slow, or stop an animal's escape. The concept is simple. The seine is placed between the animal and its escape route. As the animal makes a dash for safety it runs into the barrier. The height of the barrier will depend on the species. For birds that will be somewhat airborne by the time they reach the net, or for birds that will try to scale it, the net should be at least 6 feet (1.85 m) tall.

Static land seines can be fairly simple and easy to construct. The bottom of the net must be weighted – held down with rocks or anchored to something heavy like electrical conduit, steel pipe, or sections of rebar. The netting is then attached to posts, stakes, tripods, or a rope, drawn taut between fixed points (Fig. 18).

Mechanical nets

Active land seine

A mechanical or active seine employs the element of surprise. The net is hidden until it is triggered to spring up. As with the static seine, the netting must also be weighted at the bottom to prevent animals from scooting underneath. A long rope, much longer than the section of netting, can be weaved through the top of the material. If using the lightweight garden netting, it must be folded over a few times before weaving the rope through, making it stronger and less apt to rip when pulled.

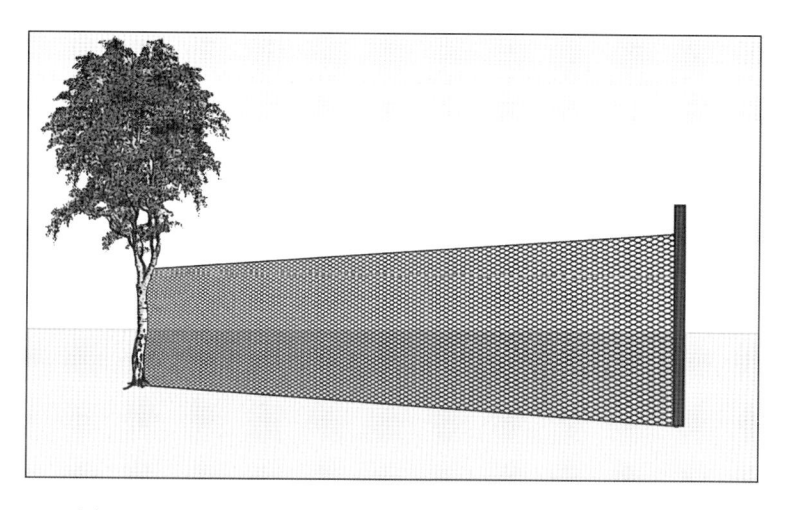

Fig. 18 A land seine is a wall of netting that acts as a barrier. Capturers use the land seine to block an animal's escape or to drive it in a particular direction. Land seines can be static or mechanical.

To prepare the seine, it is best if it is folded on top of itself, like an accordion, with the rope portion resting on the very top. If an animal is to cross over the net before it is triggered, it should be concealed with debris or covered lightly with soil. If it is to be set in shallow water the garden netting will need to be weighted down, as it tends to float.

There are various methods of launching the net. One technique involves securing the rope at one end, to a fixed object, such as a stake or tree trunk, with the other end of the rope controlled by a person, usually hidden out of sight. When the time is right, they pull the rope quickly, raising the seine.

Another method requires the use of shock cord. The end that would normally be pulled by a person can be attached to heavy-duty elastic cord, well anchored, and stretched.

The bow net and Q-net

These mechanical traps are essentially made of netting attached to a rigid frame. One half of the frame is anchored to the ground, the other half is folded back over the rigid half. It is held under the tension of one or two coiled torsion springs. Brackets, welded to the center of the stationary section, hold the release mechanism. In a sense, it is like a common mousetrap. When triggered, the top frame snaps forward, over and down. The release can be manual, using line, or electronic, using servomotors or a solenoid. Q-nets are large bow nets that are powered by large elastic cords, or bungees.

Mechanical, spring-loaded net traps come in many shapes and sizes with one or more moveable parts. Many are hand made, though there are some, such as the Moudry net traps, that are sold commercially. Proper training in the use of these nets is essential. The moving parts of these traps pose significant risk of injury or death to animals if not used properly. Because of the risks associated with these traps, they should not be used when safer options are available.

The whoosh net

The whoosh net, named after the whooshing noise it makes when triggered, uses elastic ropes or bungees to rapidly propel netting up and over target animals after they have been lured into the "catch zone" (Fig. 19). The leading corners of the net are attached to rings that slide onto poles anchored at either corner

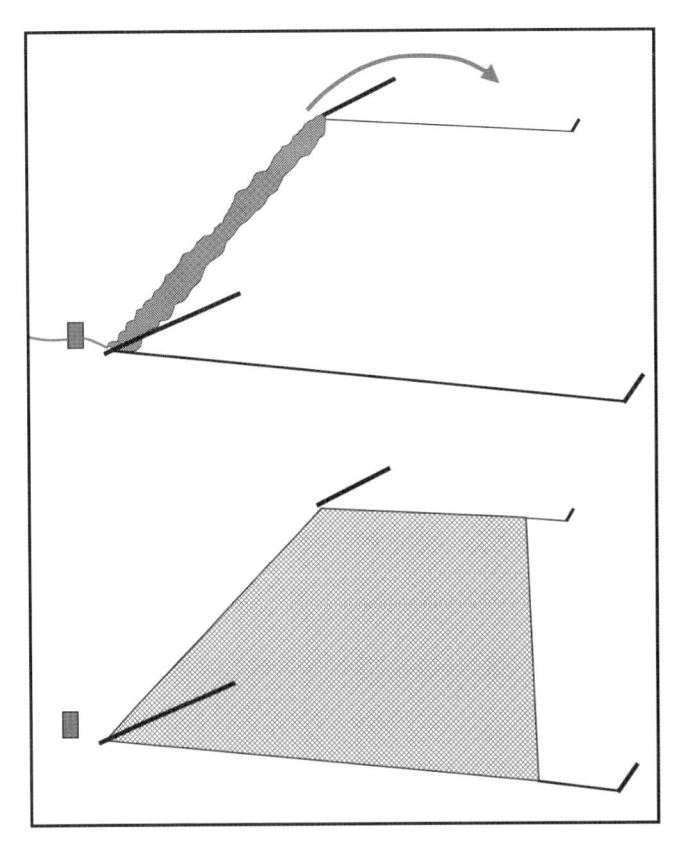

Fig. 19   The whoosh net uses elastic cords to propel netting over and on top of animals in the catch zone.

of the trap, leaning forward at about 28 degrees. The netting material is furled at the rear of the trap. The rings and net are held by a trigger mechanism at the base of the poles, under tension of the elastic cords. When triggered, the two rings shoot up and off the poles, over the catching area, and then down. The net is held snuggly against the ground by the remaining tension of the elastic rope.

Driving, funnel, and walk-in traps

As discussed earlier, barriers or "walls" made of netting can be used to control an animal's movement. The "drive" or "funnel" trap employs the use of multiple walls of netting, angled toward and sometimes into an enclosure or large cage.

Constructing the enclosure part of the trap is the first step. A circular shaped pen, as opposed to a square or rectangular one, will encourage trapped animals to travel along the perimeter, reducing the chance of them becoming injured from hitting the walls straight on, or getting "stuck" in corners. Depending on the species, and how long animals will be confined, material can be attached to the inside walls to minimize stress on the animals and further reduce the risk of them injuring themselves as they try to escape.

Extending out from the entrance to the enclosure are walls, or "wings," made of netting (Fig. 20). These panels are sometimes referred to as leaders. Animals are be driven toward the barriers and then funneled into the enclosed space. The netting

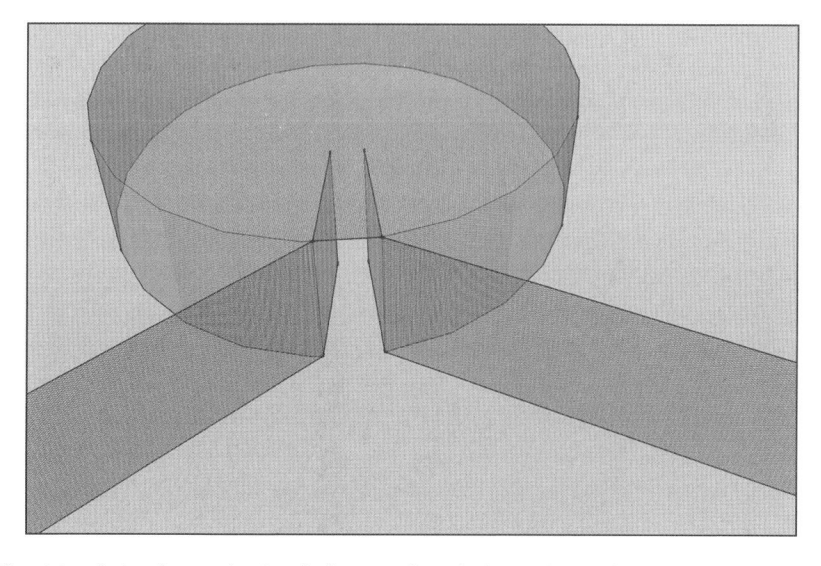

Fig. 20 Animals can be herded toward and along the baffles of a funnel trap and into the enclosure.

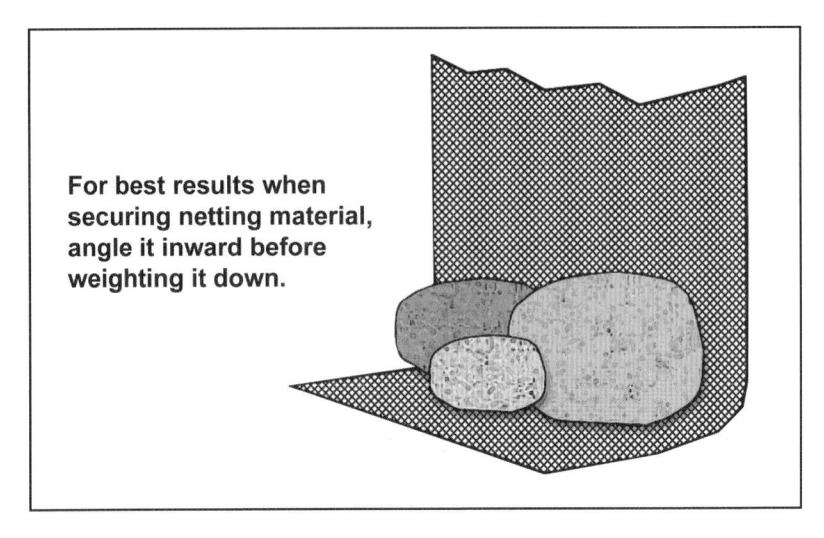

For best results when securing netting material, angle it inward before weighting it down.

Fig. 21 It is important that walls of netting are firmly secured at the base, angled inward and buried or weighted down.

of these panels should be taut and constructed of a material that will not injure or entangle the animals when they run into it. In most cases the bottom of the netting must be secured – angled inward at its base, and then weighted down (Fig. 21).

Some funnel-type traps are designed with a passage that leads into the enclosure – basically a continuation of the outside wings, or funnel. This design is often used with birds. Most birds have a difficult time figuring out how to get out once they are inside.

Herding animals toward a trap begins by applying a light amount of pressure. This could be a noise, the sight of a human, etc. Typically, this method of capture calls for a number of herders positioned in a half-circle (Fig. 22). Pressure is delivered steadily and methodically. The amount and direction of the pressure changes as needed to encourage the animal(s) to move in the right direction. Use of a long-handled net can help apply pressure and be used to capture an animal that tries to escape past the herders. A variation on this is to have the herders holding a modular section of netting, a moving "wall" that can be used to close off the animals' escape.

Some funnel-type traps are designed with a passage that leads into the enclosure – basically a continuation of the outside wings, or funnel. This design is often used with birds. Most birds have a difficult time figuring out how to get out once they are inside. These types of traps are sometimes called "walk-in" traps. One such walk-in trap, called the Ottenby, has been used to catch wading birds for decades. It is a rectangular shaped unit made of welded wire with two entrances on opposing sides (Fig. 23). Birds walk in through minute funnel entrances and

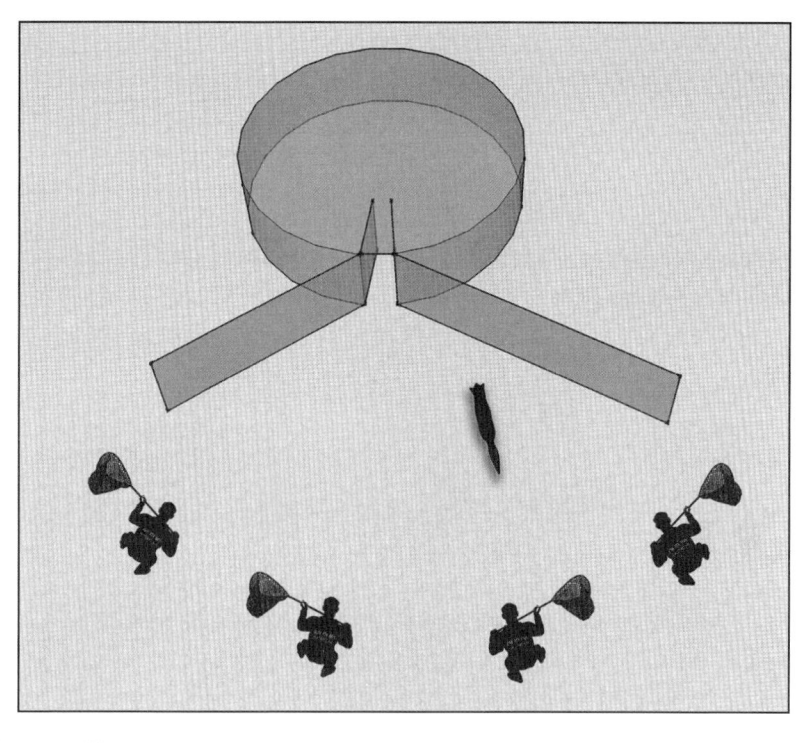

Fig. 22   When herding animals, capturers should take a half-circle or V formation.

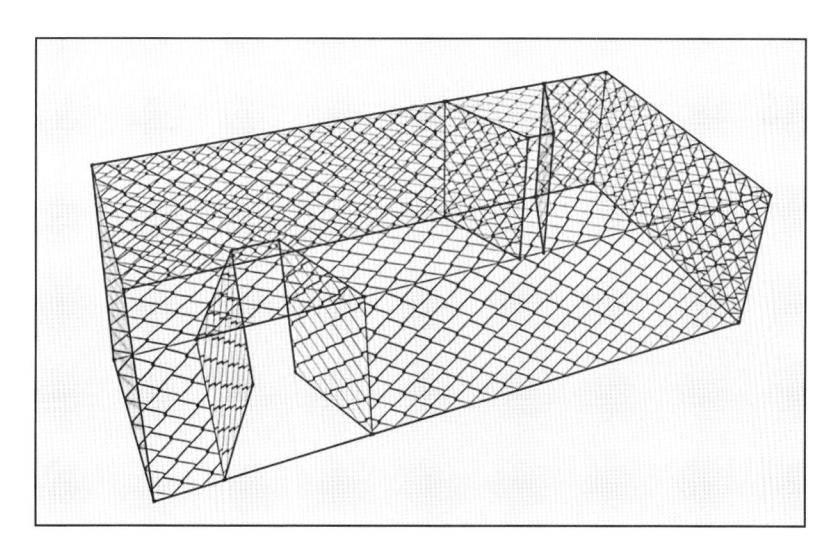

Fig. 23   An example of a rectangular walk-in trap with two entrances.

cannot figure their way out. There are many variations on the walk-in type trap – they can be square, round, cloverleaf, and heart-shaped, depending on the target species.

To have the greatest chance of success, walk-in traps should be placed in an area the animal is likely to travel to. Decoys and bait can help attract birds to the area. Natural obstacles, such as rocks, driftwood, or vegetation, can be used to guide birds in the right direction, with a leader, or panel of fencing, to help guide birds toward trap entrances. On an open shoreline, traps can be set to intercept animals foraging along the water's edge.

The dho-gaza

A dho-gaza is simply lightweight netting, like a mist net, suspended between two poles. The idea is for the netting material to be unnoticed by the animal until it's too late – until they are traveling through it. When an animal runs into the net, the material detaches and collapses on and around it.

There are two methods of using a dho-gaza. One employs a large stretch of netting and driving an animal into it. The other involves a smaller section of netting with a live lure on one side, and is typically used to trap birds of prey.

The poles that hold the netting should be lightweight and hollow so they won't injure the bird, such as fiberglass or aluminum extensions, or wooden dowels. The poles should be camouflaged with paint. The hollow poles can be "set" vertically on anchor-mounts made of metal rods that are pounded into the ground. Dowels require a wooden base, such as 2-inch (5 cm) baseboard, to which they can be glued.

The netting material needs to be strong but lightweight, and colored so that it blends in with the surroundings. The netting can be attached to the poles using clothespins, alligator clips, or paperclips secured to the poles using rubber bands for give. To ensure the netting breaks away from the clips easily, duct tape tabs are used to attach netting loop to clip (Fig. 24). If a drag is deemed necessary, a weight can be attached to the netting using elastic cord. As with any trap, the dho-gaza must not be left unattended.

Drop traps

The drop trap is one of the oldest and simplest methods of capturing animals. The design and material can vary depending on the type of animal that is being captured. The concept is basic – a net or cage, often shaped like a box, is suspended above bait or a lure. When the target animal is well underneath it, the trap is dropped, usually by a handheld mechanism. For example, a box-like cage can be tilted up on one end, held upright by a stick with a string attached (Fig. 25).

Chapter 7 67

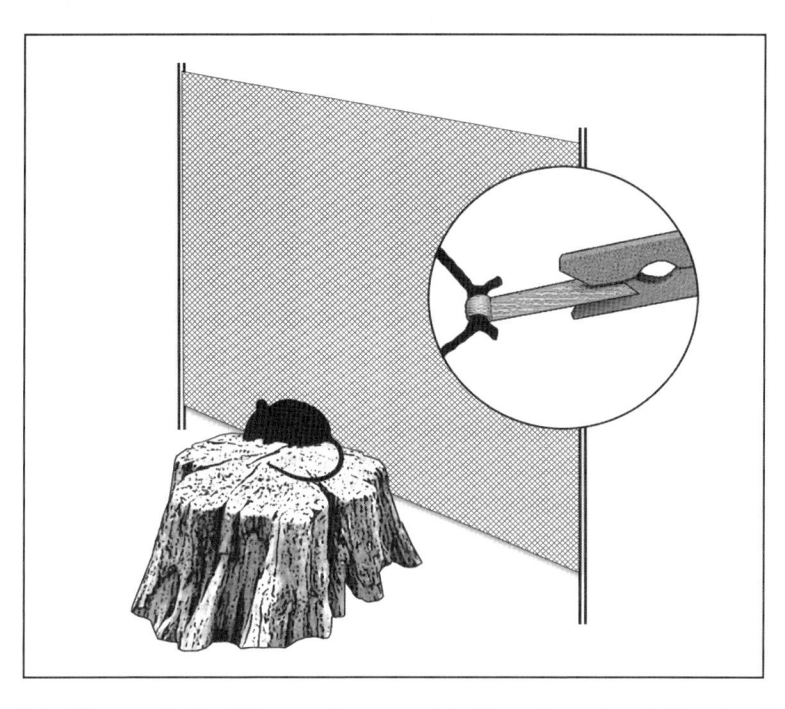

Fig. 24 The panel of netting can be secured to the poles using clothespins. Duct tape tabs ensure the netting will slip free easily.

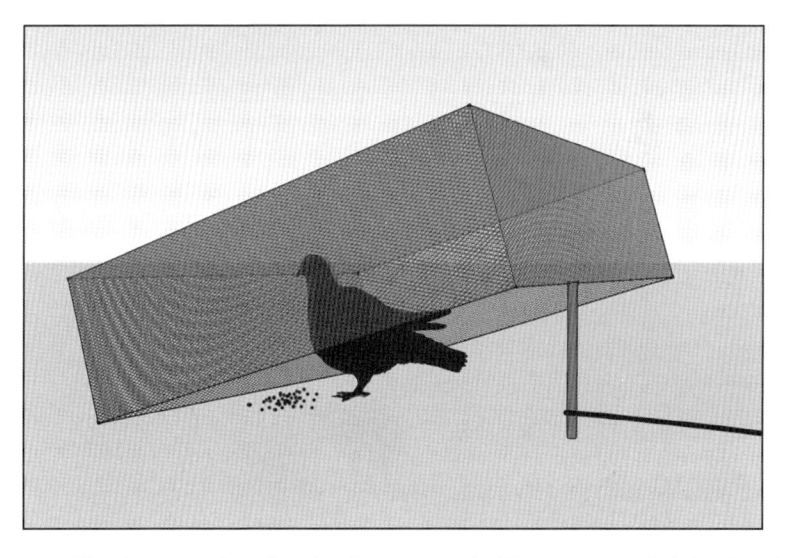

Fig. 25 The drop trap is quite simply a suspended box or cage. Bait is placed at the rear of the trap to encourage the animal deep into the catch zone.

There must be enough room for an animal to walk under and reach the bait, but not so much that it takes long to fall. It is critical that the string be drawn taut ahead of time. The trapper should be out of sight, or out of the way, holding onto the taut line. When the moment is right, the string is tugged swiftly.

For this capture method to have the greatest chance of working, the target animal needs to be deep inside the catch zone, facing away from the opening, and distracted by the bait. As with all traps, they must not be left unattended.

Cage traps

Wire cage traps come in a variety of sizes and shapes. They are sometimes referred to as humane traps. Typically they are rectangular, with openings at one end, sometimes both. The mechanics involve a lever that the animal has to step on, which causes the door to fall shut and lock into place. One of the earliest cage-type traps, the Potter trap, was designed by Jessica A. Potter in the 1920s to catch small songbirds.

While a cage trap may be the only way to catch an animal, rescuers must take precautions. As with any solid wire cage, a wild animal will be terrified beyond words and can quickly injure itself trying to escape through the wire. Traps must be monitored closely so the animal can be removed without delay. That said, periodic visits can potentially scare off the target animal and blinds may not work with wary mammals. The trap can be monitored remotely through video surveillance, or an alarm system. For example, a mini magnetic reed switch can be used to alert rescuers through a radio signal or telephone when the trap has been sprung.

Projectile-powered nets

Net guns, cannon nets, rocket nets, and net launchers use projectiles to cast a mass of netting over target animals. There are two types: pneumatic, using compressed air to launch the projectiles; and pyrotechnics, which use explosive materials. Pyrotechnic-powered devices may require compliance with certain regulations and safety policies. Owners and operators of these devices should thoroughly research the requirements for safe storage, transport, and operation.

Often used by researchers to capture a significant number of individual animals, the larger projectile-powered nets are not without their drawbacks. They are loud, and can pose significant risk to the animals and operators. Animals can be severed in two or otherwise mortally wounded by the projectiles or the netting. For the wildlife rescuer, these tools should be considered a last resort.

Typically, when setting a stationary, ground-deployed net, as opposed to a handheld one, the back edge of the net is usually secured at the rear of the catch zone with the rest of the netting folded on top of it, or just in front. The netting material can be lightly camouflaged with organic material from the surrounding area, such as leaves or seaweed.

The multiple cannon launchers are usually set behind the net. Inconspicuous markers are used to indicate the catching area so the capturers can see when the animals are in the capture zone. The timing of the cannon firing, which is done from a hard-wired electrical circuit board staged some distance away, is critical to reduce risk to any animals.

A variation of the net cannon, called a net launcher, differs in that the net is contained, loosely folded inside a bin, with two to four barrels angling out from a central chamber. Net cannons can be ground-deployed or suspended from an elevated angle pointing downward.

Net guns are similar to net cannons or launchers except that they are handheld and designed for relatively close-range use (5–10 m). The netting material is usually housed in an open canister that is centered between four angled barrels loaded with projectiles that will carry the netting up and over the target animal (Fig. 26).

Lures

Lures are used to draw animals into traps or within close range for netting. Lures can be divided into four different types: decoys, food bait, scents, and audio lures.

The term decoy refers to a real or artificial replica of an animal meant to draw the attention of another. For example, waterfowl decoys are used to successfully attract other waterfowl.

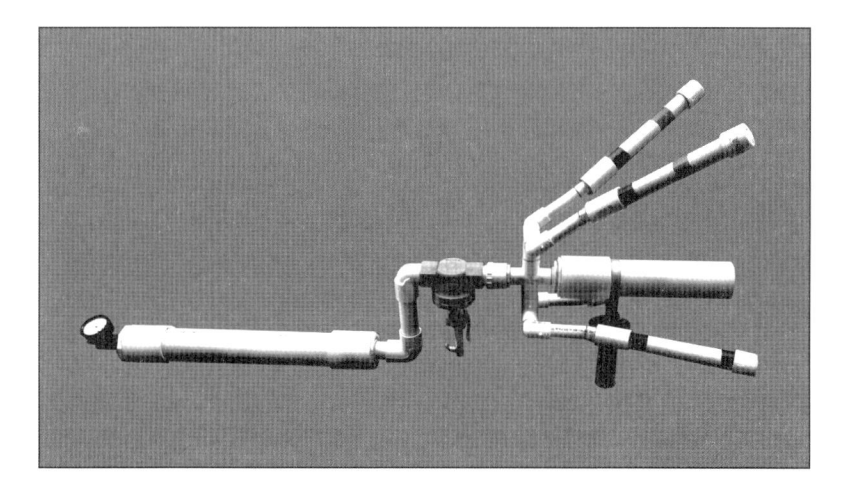

Fig. 26 An example of a handmade pneumatic net gun.

Nowadays, most audio lures can be delivered through a portable playback system. Compact-disc players, iPods, and Walkmans are examples of portable devices that can be coupled with external speakers to deliver sound. The application will vary depending on the situation. Predators, such as coyotes, can be lured using sounds similar to a dying animal. Birds can be attracted to a variety of sounds, such as begging calls, mating songs, or the sounds of a competitor.

Catchpole

A catchpole is a common tool used in domestic animal handling. It is also used to manage some mammal species. A catchpole consists of a long pole with a loop at the end – basically a snare. Most commercially available catchpoles allow the operator to control the size of the noose.

When making a catchpole, or snare pole, the cord should be rigid enough to maintain its hoop shape, opening readily when pressure is released. Thin cord is not recommended as it will not release easily and there is a risk of it cutting into an animal's skin. Nooses can also be made out of flat nylon webbing.

When a catchpole is used to control a mammal, such as a raccoon, or bobcat, the loop must never be used just around the animal's neck, but also around one of its shoulders. Canids can be manipulated with the loop solely around the neck as long as they are never lifted this way.

Overall, catchpoles are not recommended for use on wild mammals when there are other options available.

8 Capture, handling, and confinement of wild birds

Techniques for capturing wild birds

As discussed earlier, the weight and size of netting must be appropriate for the animals being captured. It must not injure them. Birds, for the most part, are lightweight and delicate. The netting material used does not need to be as heavy or weighted as that used in capturing mammals. If it is too heavy, a bird's wings can be damaged.

Another consideration is how the bird will react to the net. Unless the plan calls for scaring the birds off, nets must be held down and off to the side, not high in the air. Typically, when approaching a bird, the capturer will want to hold the net so that the hoop and bag is tucked from the bird's view, to the side or behind their frame, as inconspicuously as possible. When they are ready to lunge for the bird, the net can be brought into a position better suited to their swing, with the far end of the handle in their strongest hand.

There are a couple of different techniques for swooping on a bird with a net. If the bird is flighted, the swing should come from above (Fig. 27), carrying through so that the hoop lies flat on the ground (Fig. 28). Birds that are not flighted can be captured using a side swing motion (Fig. 29). In shallows, when chasing a diving bird that is swimming underwater, skilled netters can plunge the net straight down to intercept the bird, or scoop up slightly.

Where a bird is fluttering up against a window, the net can be gently placed over the bird. Often, the bird will fall into the bag of the net. To prevent the bird from escaping, the pole of the net can be twirled, twisting the fabric closed.

Enticing wild birds using lures

Different kinds of lures, or combinations thereof, are used to draw animals close enough to be netted or trapped. Types of lures include food, decoys, scent, and audio calls.

When enticing a bird close for netting by hand, ideally there should be at least two people involved – the person who is going to try to net the bird, and the person who is controlling the delivery of food, or bait.

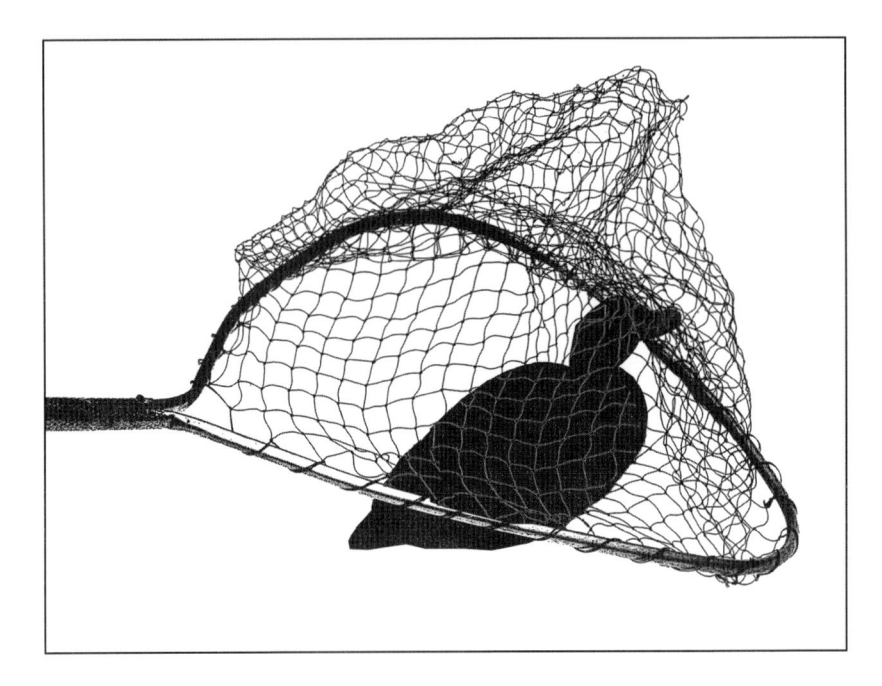

Fig. 27 When netting a flighted bird the swing should come from above.

When staging the capture, the hoop of the net can either be set down flat, so that it does not flap in the wind, or leaned up against an object, such as a car tire, a park bench, a trash can, the trunk of a tree – something that the bird will not be afraid of approaching. The netter will take a low profile, crouching, holding onto the pole of the net and waiting for the right moment.

In most cases, the person throwing the scraps of food will want to be on the same side as the netter – minimizing pressure on the animal – the perceived threat. They might also want to take a low profile, but not in such an extraordinary fashion as to alarm the bird.

While every situation will be unique, there is an art to baiting in birds. With regard to how much bait to throw, when, and where, there are a few rules.

At first, just to see if luring will work, the person baiting will want to toss only a few tidbits. This is just to see how the bird reacts. If it is interested and appears as though it will come closer, they will want to continue.

The next phase is to bring the bird in closer with tosses of irresistible treats. The food should be in small pieces and delivered in relatively small amounts so that the bird does not become satiated and lose interest. Sometimes the target bird will come in close enough for netting simply by tossing the bait closer and closer to the net, aiming for just in front of the hoop (Fig. 30).

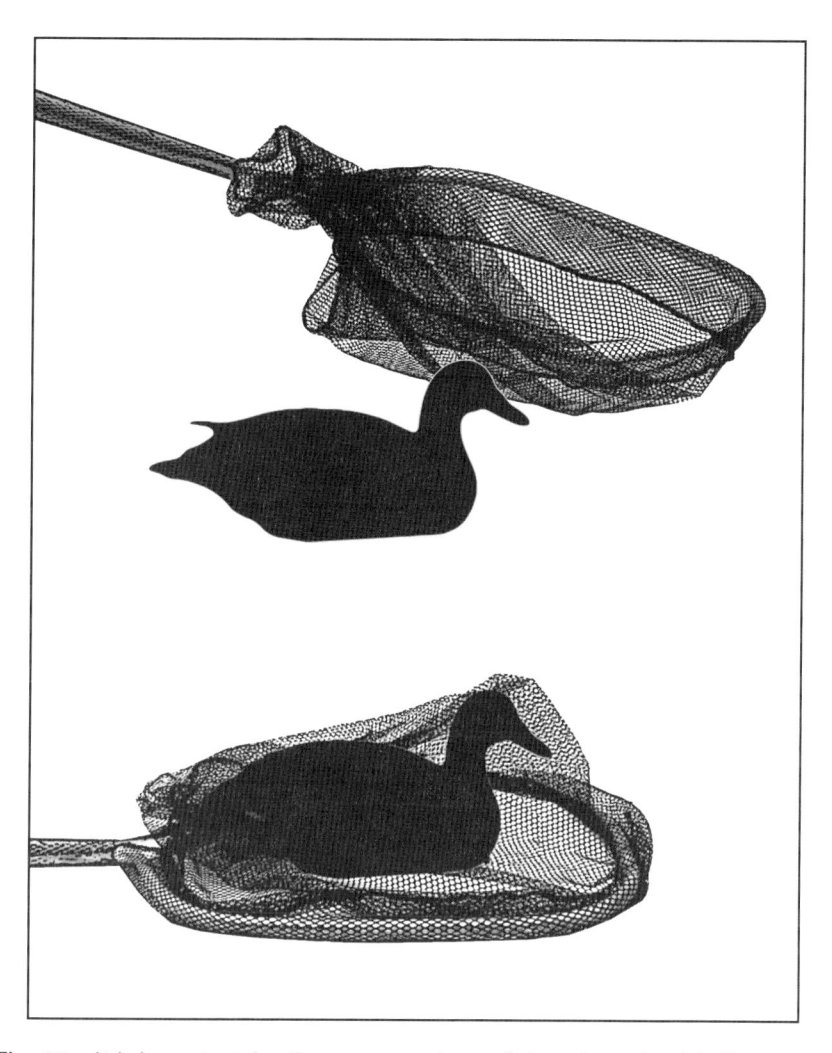

Fig. 28 It is important for the netter to always follow through with their swing of the net so that the hoop lies flat on the ground, reducing the chance of the animal escaping.

To get a shyer bird to come closer, a large pile of crumbled food might need to be dropped just in front of the net. After placing the mound of food, the rescuer should walk away, past the netter (Fig. 31).

Often, baiting will draw in other birds. This can be advantageous. An excited feeding frenzy can help focus the target animal's attention more on the food and not the humans. Sometimes, though, a disabled bird will remain on the fringe of a group, or may be chased away by more dominant birds. In this instance, these

Fig. 29 When netting birds that are not flighted, the capturer can use a lateral swing of the net.

Fig. 30 Bait is placed just in front of the hoop. When the target bird is within the capture zone the netter waits for it to put its head down before flipping the hoop over its body.

Fig. 30 (*continued*)

Fig. 31 Using a large pile of crumbled food to attract a shy bird.

animals should be distracted with food placed to the side, away from the net, so the one bird can be drawn in closer.

Regarding the type of food to use, consider the bird's normal diet and what it would probably be attracted to. The environmental conditions must also be taken into account – crisps in a heavy wind will simply not work.

Urbanized gulls will eat almost anything, from chips to scraps of meat, leaning toward high-fat foods, such as cheese-filled hot dogs and Big Scoop Fritos.

Non-urbanized gulls are not usually drawn in by junk food but will readily devour fish, especially the rich fatty parts. Pelicans can be lured in with baitfish, scraps of fish, and sometimes slices of bologna. Ducks, coots, and geese can be attracted with grain or breadcrumbs.

For shyer species, such as raptors, herons, and egrets, and many shorebird species, a mechanical trap or snare-type trap may prove a better option. They can be drawn in using live bait. The movement of small live fish in a container, a wriggling worm on a plate, or a darting rodent can be irresistible to a hungry bird.

The Bartos trap

The Bartos trap is typically used to capture raptors that are hesitant to venture to the ground. Designed by Robert Bartos, this live trap can be suspended at almost any height, either indoors or out. Its shape is funnel-like, with a wide-mouthed opening at the top and a small compartment for the live lure at the very bottom. The idea is that a hungry raptor will be enticed into the trap and onto the perch, which triggers the top to quickly flip shut, like a bow net (Fig. 32). This style of trap allows a raptor to be secured in a compartment made of soft netting, rather than wire. As with all traps, they must not be left unattended and animals must be removed without delay. If live bait is being used, it must also be tended to.

Snare-type traps

Small slipknots can be used to entangle the feet and legs of birds. It is extremely important to get the selection of weight and strength of the material, such as monofilament fishing line, just right. There must be no chance the line will break when the bird is snared. The line must also be able to keep its hooped shape, but not be so heavy that it wants to open back up after it closes on the bird's leg or its digits. No snare trap should ever be left unattended and animals must be freed from the snares quickly. See Appendix 4 for instructions for creating nooses and tying them to wire mats.

Bal-chatri

The bal-chatri is basically a wire cage containing a live lure. The outside of the cage is adorned with multiple monofilament nooses (Fig. 33). Additional longer nooses can be attached near the base, extending beyond the cage for birds that

Fig. 32 The Bartos trap.

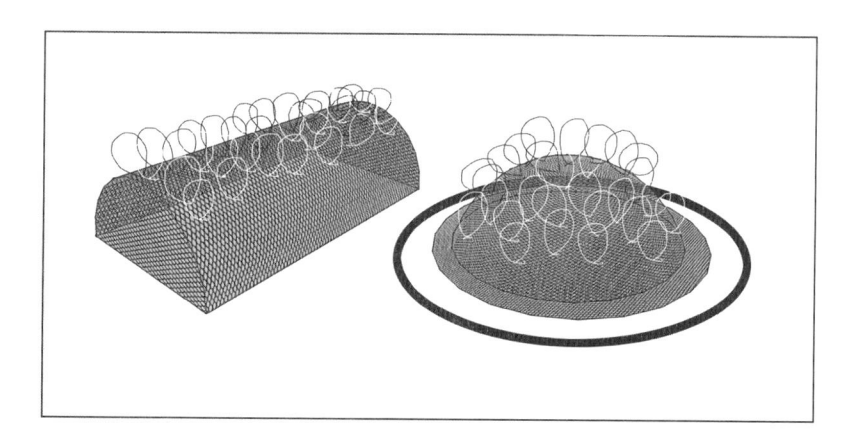

Fig. 33 Bal-chatri traps can vary in size and shape.

may be resistant to perching on top of it. The traps should be painted to blend in with the natural surroundings.

The size and shape of the cage and strength of the line used for the nooses will vary depending on the target species. A wider, rather than narrow, cage will allow the bait animal to move about more, which is apt to get the attention of the bird of prey more quickly. Rodents tend to travel against a surface and they like to take up in corners. Round designs will encourage rodents to keep moving.

Generally, the nooses are secured to the trap using a square knot or clinch knot. The noose itself is made using an overhand knot, although a running slipknot can also be used. Nooses are attached at the junction of the wire squares.

For this method of trapping to be successful it is important for the nooses to stay open and stand vertically. To encourage this, the anchor knots can be rotated until the noose "stands up." Applying a drop of plastic cement can also help the nooses stay in an erect position.

When the trap is set, it must be anchored to something the bird cannot fly away with, such as a weight. It is very important to attach the weight to the cage using an elastic cord as a shock absorber. This will reduce the risk of injury to the bird and lessen stress on the nooses when they are pulled.

The phai trap

The phai trap usually consists of a cage or small area containing live bait that is surrounded by large nooses. It can be constructed in various shapes (Fig. 34).

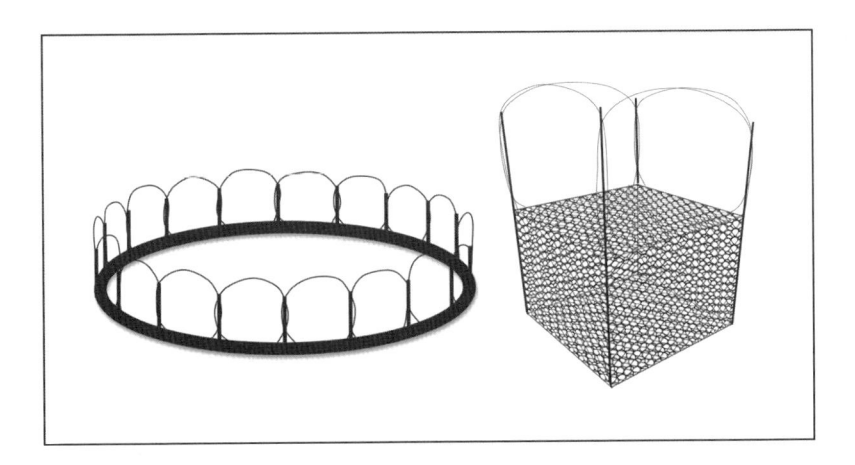

Fig. 34   Variations of the phai trap.

The "box" version uses a square cage. At each corner, lightweight but somewhat stiff twigs or rubber rods are placed vertically, to which nooses are attached and held open. The strength of the monofilament line will depend of the size and strength of the target species. As with the bal-chatri and other traps, the phai must be grounded with sufficient weight using an elastic cord to reduce shock.

The ring-shaped version can be constructed out of a rubber hose or rope that is cut and pieced together to form a hoop. The nooses are then tied through or onto the hoop in such a way that each noose overlaps the next.

A variation of the ring version uses a dowel that is drilled with small holes to hold a row of nooses. Nooses can be fixed using a clove hitch. Plastic glue can be injected into the holes to fix them permanently, and can also be used to help the nooses "stand up." Eyebolts are affixed to one or both ends of the dowel, to which a shock absorbing elastic cord is used to attach a weight.

Noose carpets

Noose carpets are weighted pieces of material festooned with multiple nooses. The material is usually made of hardware cloth but large nylon netting has been used successfully as it will mold to a surface, such as a rock. As with all the noose traps, the noose carpet should be fixed with a weight using a shock absorbing elastic cord, such as a bungee or surgical tubing.

Noose carpets are extremely versatile. They can be placed where the target bird is likely to land or bait can be used to encourage birds to walk over the mats. Birds can also be herded over them.

The single snare

Another method of snaring a target bird is with a single noose. The slipknot is set open, raised slightly – just enough for the bird to step into it (Fig. 35). Bait is placed in or around the snare, depending on the setting.

The other end of the line, the one that is held by the capturer, should be tethered securely to their wrist so it cannot slip free. The weight of the line is critical – it must be able to withstand a sharp tug and the bird's struggle without snapping. Gulls can be captured using 50 lb test while a higher gauge, such as 80 lb test, should be used for larger birds like pelicans. There are both advantages and disadvantages to large-gauge monofilament line. While thicker line tends to hold its shape once it is set, it has a tendency to want to open back up.

In preparing to snare the bird, the line should be drawn up to reduce slack; too much line and the bird might be alerted to its movement. Once everything

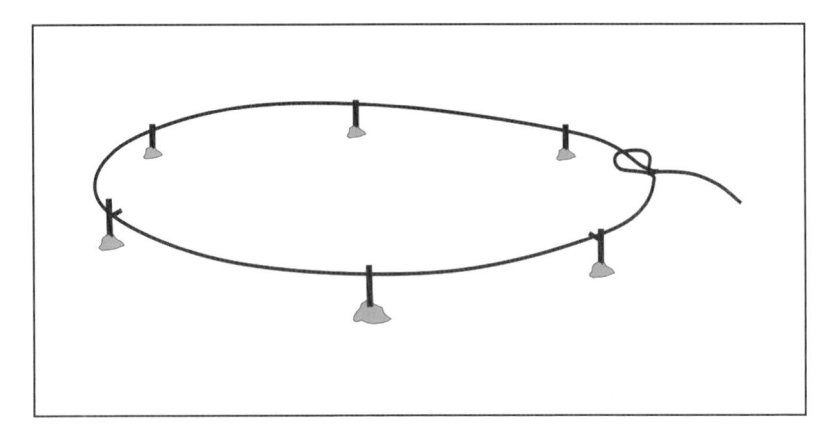

Fig. 35 A single monofilament snare can be raised slightly off the ground with small twigs.

is in place the capturer should keep fairly still until the bird is in the snare and preoccupied with the bait. They then quickly pull back on the line.

Although one person can perform this technique, it is best if there is an assistant to help place bait, ward off non-target animals, and catch up the bird once it is snared.

Anecdote: The rescue of the elusive beer-can-collared-gull

It was late September when the first Internet post appeared. Someone near San Francisco, California, had observed an adult western gull with a beer can around its neck. This was obviously someone's malicious doing. It seemed an isolated incident and the bird was not reported again.

Then, in November, another sighting was reported, but this time it was a juvenile western gull. In spite of concern that reaching out through the news media might somehow encourage people to try to capture the birds themselves, we went ahead with a press release. We were specific with our message that any attempts to catch the birds would only make their eventual capture more difficult. We asked instead that all sightings be reported through one dedicated number or e-mail address.

The reports started flowing in. Soon we had up to five different gulls that had Budweiser beer cans around their necks. The sightings were spread out across the Bay Area from the Fremont landfill to quiet Stinson Beach, and south to Pillar Point Harbor. A vast area.

We took the information from the most current sighting and drove the two hours to Half Moon Bay. While we rescued a flighted gull that was horribly tangled in a crab trap, we found no collared gull.

We drove north toward San Francisco, stopping to scan popular loafing sites, parking lots, and piers, but saw nothing. Finally, through our paging system we received a call from a woman who had just observed one of the can-collared gulls at Lake Merced, just minutes away.

There it was, standing among other gulls on the railing of a wide cement bridge used to link portions of the park. There were no vehicles allowed on the bridge, but pedestrian traffic was high.

Image courtesy International Bird Rescue, online at www.Bird-Rescue.org.

The bird looked good: he was bright and alert and other than a few mussed up feathers, he seemed all right for the time being. The issue, though, was that the can was preventing the bird from preening and bathing and would eventually cause it health complications. The can separated the bird's feathers around its neck, allowing cold water and wind to reach its skin and causing it to lose body heat. Eventually, this would cause the bird to become debilitated and die.

With the gull used to people feeding it at the lake, our first approach was to try to bait it close enough for netting. That did not work. We could have captured thirty other gulls, but this particular bird was shy. Understandably so – he had been lured in and captured once already.

Our next approach was a drop trap. Using PVC and garden netting we set a large, rope-anchored (handheld) drop trap and placed delicious Fritos Chips and cheese-filled hot dog scraps at the back of the trap. He did not go for that either. This may have eventually worked but the public park was no place to leave a contraption for the birds to get used to.

The next method was the leg snare pole. Using 50 lb test monofilament line and a long bamboo stick, we went down onto the lake's shoreline and baited in the birds. That almost worked.

With all the media coverage, we were getting a tremendous amount of attention and enough donations for us to purchase a Coda NetLauncher to use. Among the many calls and e-mails, we were contacted by a woman who said she regularly visited the park and fed the gulls and was very concerned for this poor animal. She went on to describe how she had been careful in developing a relationship with this gull and had gotten to the point of being able to place food on the railing of the bridge and then she'd step back while the bird came up to eat the morsel. She would place another piece and step back, and so on.

It had been nearly two weeks since we first tried to capture this particular gull. One morning, not too long after hearing from the woman about her relationship with the gull, I woke with an idea that involved twigs and bubble gum. Excited by the new idea, we made our fourth trip to Lake Merced. Unfortunately, the gull was nowhere to be found. We waited. Before noon, the sun broke through the marine layer. Finally, he showed up.

While we had one of our volunteer rescuers keeping him occupied with tosses of chips, Duane and I set up the snares. We were going to use the bird's habit of standing on the railing, its familiarity with the woman who had been feeding it regularly, its social status among the others, and its love of cheese-filled wieners to get it to step into a snare.

With two 10 foot (3 m) long sections of monofilament we created nooses at each end – one for the operator's wrist, one for the leg snare. With the involvement of my electrician-husband, the bubblegum was replaced with black vinyl mastic insulating putty. The putty was used to hold very small crooked twigs to elevate the snare about a half inch (1.25 cm) off the cement. They were placed on the surface of the cement railing and mounted on the vertical sides as well. The snares were set apart, leaving about an 8 inch (20 cm) space in which to set the bait. Once we were all ready, Duane and I stepped back and slightly to the side to allow the placement of the food.

Initially, because the target bird was not near the trap when the bait was set out, we had trouble with other gulls trying to get at the meal. We also had to shoo an adult western away, allowing the can-collared bird to assume its spot in the pecking order. From the time we first set the trap, we had him stepping in and around it within five minutes. Thinking I had him, I pulled the noose. Twice. The wonderful thing about using this method was that the gulls were oblivious to our failed attempts – they were quite used to stepping on fishing line and seeing it fly in the wind.

Soon enough we had the nooses set again. I handed my line to Duane. Within a few seconds the collared gull and another gull were on the bait. Duane pulled and snagged the collared-gull's leg. I quickly ran in to secure the bird.

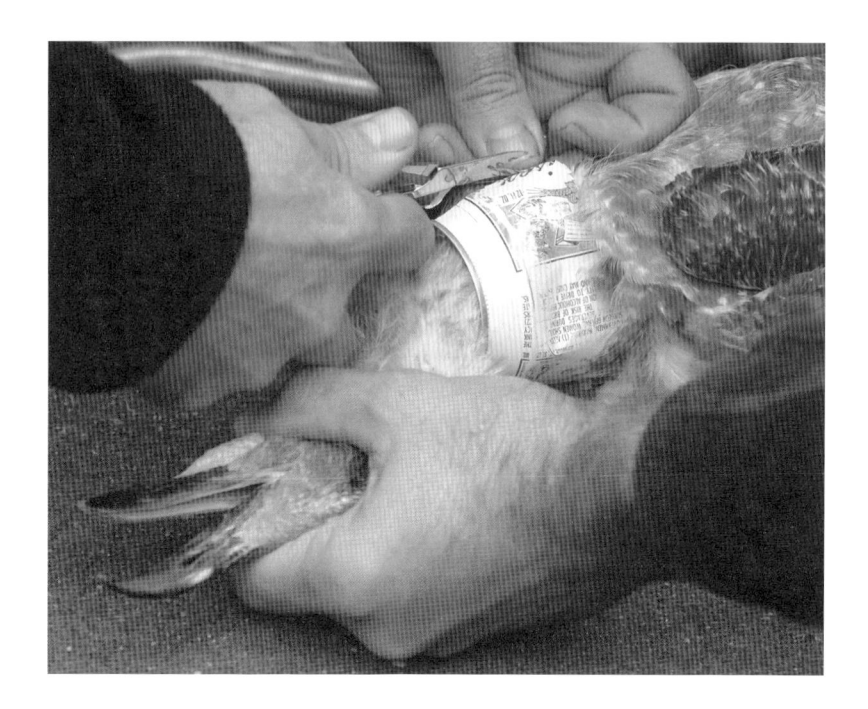

Using metal cutting shears, Duane cut the can off the poor bird. We examined him closely and felt he was in very good shape. We'd previously discussed options with International Bird Rescue, the center that would have received him if he needed treatment, and agreed that if he looked good and was in good condition, good weight, he should be released. We set the gull on the railing and let go – he took off into the wind and landed in the lake some distance away, where he took a very long bath. Later, as we were packing up the car to head home, we saw him again, setting in a flock of other young gulls, perhaps hoping for another handout.

Leg snare pole

The leg snare pole differs from the single snare described earlier in that a long stick is used to extend the reach of the capturer. While one end of the line is still tethered to their wrist, the noose end is secured to a pole. This gives the capturer the ability to flick the line upward or to the side, quite quickly, with little effort or body movement.

There are a number of things one can use for the pole. It can be a modified fishing rod, a slender stick of bamboo, or a wooden dowel, ideally something natural the bird is used to seeing in its environment. The pole does not have to be strong. Because the line is tethered to the capturer, the pole just needs to be strong enough to close the slipknot when flicked.

The line, however, must be strong enough to withstand the initial tug as well as the bird's struggle. Medium-sized birds, such as gulls, have been captured using 50 lb test, while larger line is used for pelicans. In the end, it is best to err on too heavy a line rather than having it snap during the rescue attempt.

In preparing the leg snare, the noose can be attached to the tip of the stick using tape or a cable tie. Again, it is not important that it stays attached once the bird is snared because the other end is secured to the capturer's wrist.

The snare is then placed on the ground where the bird will feel safest. It can be raised up off the ground slightly using twigs if the capturer is planning to flick sideways, or it can be lightly buried in sand or with a dusting of soil if they intend to flick upward.

Just like the single snare method, this method usually requires the assistance of a second rescuer. It is their job to draw the bird toward and into the snare by placing or tossing bait. They will also be responsible for running in and catching up the bird after it is snared.

Swan hook

The swan hook is just that – a hook used to capture and control swans and related birds. It consists of a narrow shepherd's crook at the end of a long, lightweight pole. The hook is placed gently around the bird's neck, where it acts like a collar, giving rescuers time to gain control of the wings and feet.

Pit traps

Pit traps, or "dig-ins," have been used to successfully capture large birds of prey, such as eagles, condors, and vultures. A pit trap is constructed by digging a large cavity, large enough to hold a person. Bait is placed nearby. The idea is that the target bird will land near the bait and the person will grab hold of the bird's legs.

There are quite a few variations to this method. One calls for the person's body to be "buried" except for their head, shoulders, and arms, which are camouflaged. Larger, more involved, pits are constructed using plywood and support beams to prevent the dirt walls from falling in. They can be made with trap doors and openings and fitted with camouflaged "baskets," allowing a capturer to wait in a seated position with their head in the "basket" and their arms coiled and ready to grab the bird's legs. The lure, such as a fresh carcass, is placed about a foot

(30 cm) away. It is important that the birds do not see a human enter the trap. Therefore, it should be manned just before sunrise.

Although this method takes time to construct and dismantle it is extremely effective, and probably the safest of all methods for capturing eagles, vultures and the like.

Mist nets

Mist nets are large, practically invisible "walls" of netting that are widely used in bird research and banding programs. Because there is no way to control what bird is caught, the risk of catching numerous non-target birds is too high for mist nets to be a tool of choice for most rescues. It is also possible for birds to be injured when they are removed from the net.

When birds fly into the netting they become tangled in the lightweight mesh. It requires patience and skill to remove a bird safely. During the process a bird's eyes should be covered to reduce visual stressors. Handlers must work methodically to remove the animal, beginning by softly immobilizing its feet and wings and then determining what side of the net the bird flew into. In most cases, the tail and one wing are extracted next. If the bird appears hopelessly tangled rescuers should not hesitate to begin cutting single strands until it is freed.

A word of caution: mist nets should never be left unattended and must never be set in rainy conditions as birds will be vulnerable to hypothermia.

On the water

As a rule, water rescues are more difficult and come with obvious risk. That said, sometimes they are the only plausible means of recovering an infirm bird. The following paragraphs detail capture strategies that have been used to successfully capture birds on the water.

With regard to an appropriate vessel, the lower the profile the better. Shallow aluminum skiffs have been used with success but rigid inflatable boats (RIBs) are ideal for most rescues. A dark colored boat is preferable.

When approaching a marine bird that is settled on a rock in the water – for example, an auk on a breakwater – the bird should be approached straight on. The speed of the approach will depend on the animal and the type of boat. Too slow, and a flighty bird will dive into the water; too fast, and there is risk of damaging the boat and its occupants.

Ideally, begin your direct approach from a good distance away, beyond the bird's safety threshold – in other words, where the bird does not perceive you as

Fig. 36   When attempting to capture a bird onshore, the boat should approach straight on.

a threat. Classically, two capturers with long-handled nets will be positioned in a low profile at the bow, nets held to the outside, anticipating the bird's flight to the water (Fig. 36). If the bird is weak or immobile it can be netted from above and gently scooped into the sock of the net. Safety precautions should include warnings to others in the boat about the net handles and how they may be flaying wildly at any given moment.

When attempting to capture birds that are floating on the water, consider the species and how it will react to your approach – it will either dive or try to take to the air. The following paragraphs examine various methods for capturing different species.

Cormorants, grebes, loons, alcids, and impaired sea ducks are likelier to dive in response to your approach than take flight. This makes them extremely difficult to capture on the water with nets, especially during the day. When approaching these birds by boat, the vessel should advance directly on the bird, relatively slowly, and *with* the wind to reduce the slapping of chop against the hull. While these can be extremely difficult captures, the following methods have been successful.

In the scoop method, one or two netters can be stationed at the bow of the boat with the long-handled net(s) out to the side, held slightly behind (Fig. 37). As the boat closes in, the operator should try to place the bird just off center, on the side

Fig. 37 When attempting to capture a bird on the water, the boat should approach just to the side of the bird.

of the strongest netter. When the animal is within reach, the netter will attempt to scoop the bird up and out of the water with a somewhat downward plunge – not over the top of the bird, but, anticipating its direction of travel underwater, slightly in front of and below it. This maneuver requires a tremendous amount of upper body strength as well as accuracy. Nets used in this fashion should be made of large mesh to reduce drag, the socks must be deep enough that the birds do not climb out, and the handles must be lightweight but extremely strong.

There is another method for netting diving birds, one that requires much less upper body strength. The approach is similar in that the boat should move straight on to the bird, but extremely slowly. The netter will be leaning over the bow, holding the long-handled net just under the surface of the water. When the bird is directly over the hoop, the net is lifted quickly.

Floating gill nets

These are large, neutrally buoyant nets that are set afloat in fairly calm water. The idea is to herd birds over the netting and then frighten them into diving into and becoming entangled in the mesh. This method of capture is not recommended as the risk of injury or mortality is too great. These nets can also be harmful to other animals that might encounter them.

Floating barriers and submersible pens

Grebes, loons, scoters, and alcids can be very difficult to capture on the water, especially during the day; when approached they will either dive or scuttle away. The following are concepts for floating devices that can be used to help capture these elusive species.

Floating barriers can be made out of a variety of materials. Garden fencing or netting can be used to make floating fence-like panels, which can be suspended using custom designed floating "posts." These panels can be used to guide birds in a particular direction.

The "posts" can be constructed out of aluminum electrical conduit or PVC pipe. The pipes are suspended vertically by slipping them through the center of a custom made block-shaped Styrofoam buoy. A set of gym weights must be attached to the base of each one to stabilize them in a vertical position. The size and shape of the buoys will depend on how much they have to keep afloat.

Another possibility, especially for long stretches of barriers, is to use containment boom as the floating base on which to tether wire garden fencing. Weights will have to be applied to the base of the fencing to keep it suspended vertically.

The size of screen or mesh is critical. If using rigid wire garden fencing, the openings must not be so small that the bird will get its head stuck if it tries to go through. The bird must be able to pull its head back out easily. It must also not be so large that the bird can manage its body part way through, only to become stuck.

When using any type of fabric netting material, the weave needs to be quite a bit smaller than the bird's head so there is no chance it will get tangled. When the openings are large, birds can become hopelessly entangled and possibly drown.

Driving a bird toward and along the floating barriers can be achieved with a manned canoe or remote controlled boat. The pressure should keep the bird moving slowly along the border, but not be so great that the bird panics and dives under the barrier.

A submersible pen is another concept worth considering. This is a circular wire pen made out of wire garden fencing, approximately 2.5–3 feet (75–90 cm) high, and PVC pipe, pieced together to shape the round enclosure. The PVC frame attaches to the outside of the netting, about 6 inches (15 cm) from its base. Netting is stretched across the bottom and secured using cable ties. Floats can be attached as needed to increase buoyancy. Three or four small guide wires should

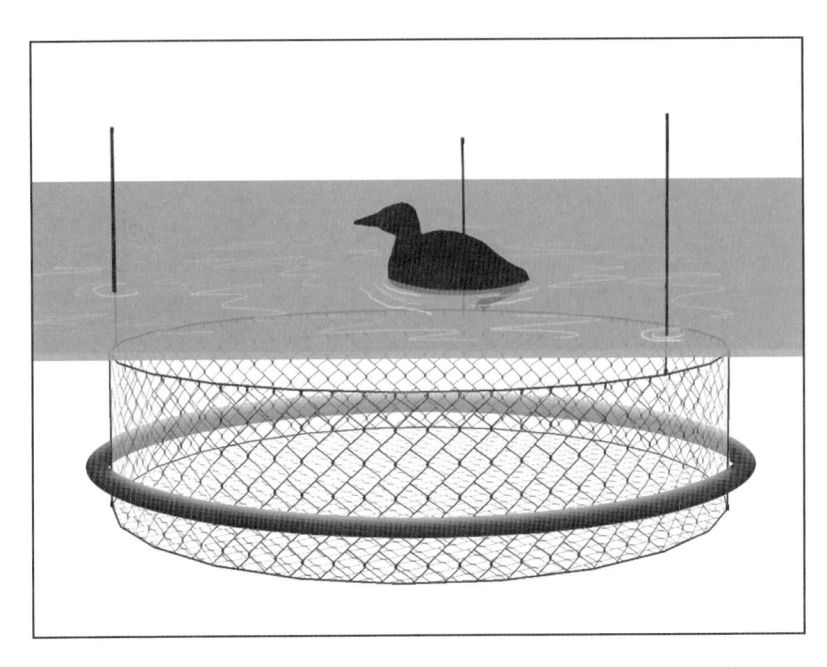

Fig. 38   A submersible pen can be used to capture certain species in shallow water.

be attached to the fencing so that when the trap is submerged, rescuers can tell when a bird is centered above the trap (Fig. 38).

The pen can be submerged using a rope attached to its base. The rope is fed through a pulley that is attached to an anchor. In shallow water a set of gym weights can be used. A person on a boat or on land applies tension until the pen is at least 1 foot (30 cm) under the surface of the water with only the guide wires visible.

Using a remote controlled boat or other means of applying pressure, the bird is encouraged toward the trap. Once the bird is in the center of the trap, tension is released and the cage pops up to the surface.

Spotlighting

Spotlighting involves the use of high-powered spotlights to find and stalk individual birds at night. When developing a plan to work under the cover of darkness, rescuers must consider existing environmental conditions. Streetlights, the moon, or glow from city lights might make conditions too bright. Clear moonless nights are optimal. Dark clothing is a must.

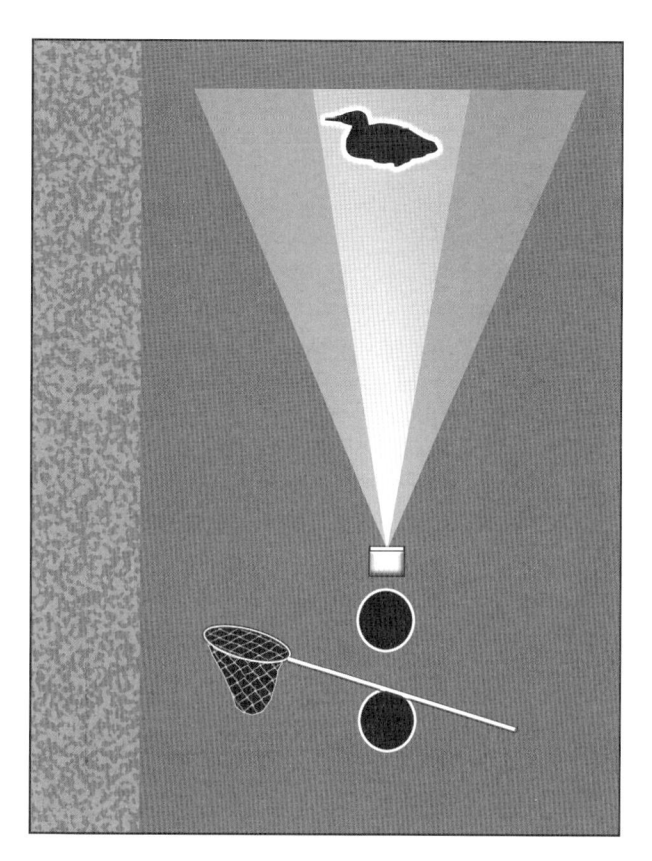

Fig. 39 When using a spotlight to sneak up on and capture birds, it is important for the netter to stay just behind the spotter, making sure the net stays out of the beam of light.

As for the technique, one person should be designated the spotter. All others should stay behind the spotter and out of any stray light cast by the spotlight. The spotter will want to scan an area quickly but methodically with the beam of light. When a target bird is located, the spotter keeps the light fixed on the animal while quietly approaching. The capturers will stay just behind the spotter, waiting to net the bird (Fig. 39).

While some species will tolerate spotlighting, others will take flight immediately. Alcids, cormorants, petrels, shearwaters, coots, scoters, grebes, loons, and some shorebirds have been successfully captured using this method. Most dabbling ducks, pelicans, and waders will tend to scare off the moment they see the light.

Spotlighting birds on the water, from a boat, is an excellent strategy for capturing loons, scoters, and alcids. The technique is similar to stalking birds on land. Ideally, the person in charge of spotlighting will take a low profile position at

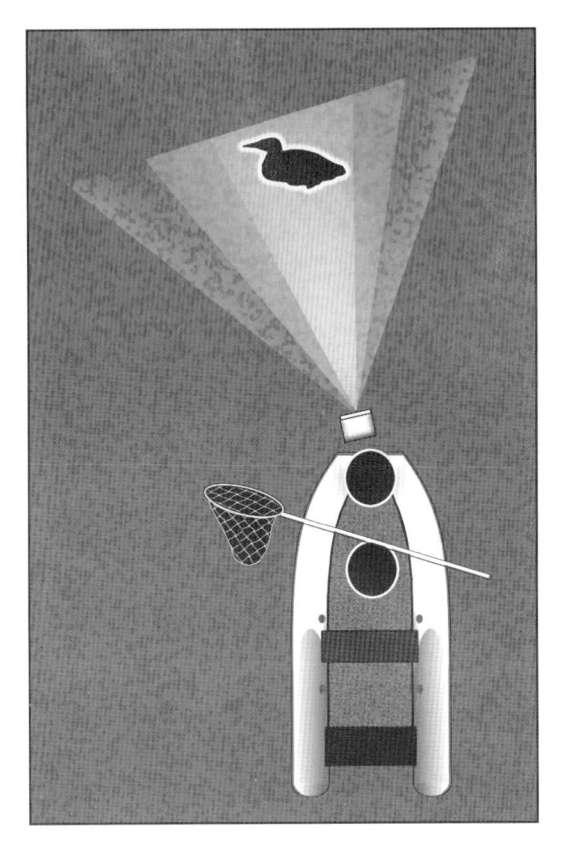

Fig. 40 When spotlighting from a boat, typically the spotter is stationed at the bow of the boat with one or more netters behind them.

the bow of the boat, scanning for birds. Once a bird is identified, they will want to keep a steady beam of light on it. As they approach netters will position themselves behind or to the side of the spotter, making sure to keep the long-handled net from being lit up by any stray light (Fig. 40). As they close in, the captain will want to position the boat to one side of the bird. When the bird is within range, the netter will make a swift attempt to scoop the bird from the water.

Special circumstances and particular methods

Hummingbirds

In most situations, rescuers will not need to provide nutritional support to any of the animals they assist. Hummingbirds, however, are the exception. With their

Fig. 41 An illustration of the proper placement of a hummingbird's bill slightly inside the opening.

fast metabolism, hummingbirds require a constant supply of energy during the day. A simple emergency elixir can supply a hummingbird with the necessary fuel it requires. It is a sugar and water mixture of one part purified water to four parts sugar. The fluid should be warmed slightly. The bird, too, must be warm in order to receive it.

The elixir can be delivered through an eyedropper or a syringe. The hummingbird can be offered the fluid by positioning the tip of the syringe or eyedropper slightly over the bird's bill – inserting it slightly (Fig. 41). The fluid must never be squeezed out – the bird will take up the fluid on its own. It is extremely important that no sugar-water gets on the bird's feathers.

Loons (Gaviiformes)

Loons are built for life on water, not land. They are rarely found on land except when injured or ill, or during the breeding season. However, they have been known to travel across mudflats or up and over levies to escape capture.

Loons can be found in freshwater and saltwater habitats. With their boat-like shape, they are excellent swimmers, using their feet to propel themselves through the water. Ill or injured loons are likely to seek the shelter of a harbor, especially in daylight hours.

To take flight, most species must skuttle across the water into the wind to become airborne. Only the red-throated loon is able to take off from land. Because of this, loons can become trapped in small, fenced-in ponds.

When approached on the water, a loon will usually dive to escape capture. On land, loons beached high on the shoreline will have a tendency to freeze rather than dart toward the water. Hoping to go unnoticed a loon may lower its head and outstretch its neck, blending in with its environment as best it can. This response will depend on how concealed it feels, the proximity of the predator, and the direction in which the predator is moving. If the bird believes it can make it to the water before being captured, it will try. If it believes the predator is too close but that it may travel past, linearly, the loon may opt to stay put.

Suggested methods for capturing loons: spotlighting; land seine; stalking in daylight.

Grebes (Podicipediformes)

Like loons, grebes can be found in fresh and salt water. Shaped like little boats, they are awkward on land, beaching only when ill or injured. When debilitated, they may seek out the protected waters of a harbor and may favor certain haul-outs, such as boat launching ramps, or sandy beaches. Along an ocean coastline, debilitated grebes seem to gravitate to freshwater outlets, beaching near the mouths of rivers and streams.

Grebes seem particularly sensitive to sound. They are easily disturbed by calls from other birds, such as gulls, or loud noises.

When pursued on water, grebes will tend to dive rather than fly, as they require a great deal of distance to become airborne, especially if there is no wind.

Spotlighting, both on the water and on land, has proven an extremely successful method of capturing Clark's and western grebes. On the water they have been known to actually approach the light.

Suggested methods for capturing grebes: spotlighting; land seine; stalking in daylight; floating barriers and submersible pens. They can be lured using decoys and vocalizations of conspecifics.

Rails and coots (Rallidae)

The American coot (*Fulica americana*) is a member of the rail family. Rails can be very tricky to capture and confine. They are very observant, with a seemingly calculating wit. When cornered, they will not hesitate to travel straight toward a human, even glancing off a person's body if it means their escape.

Unlike other rail species, the American coot is gregarious, often seen in large clans. When fleeing from danger they tend to stay grouped, at least in pairs. If they disperse – for example, during a capture attempt – the group will gather back together soon after. Coots tend to show site fidelity for foraging and loafing sites, visiting the same places every day at about the same time. These behavioral patterns can be useful in planning a capture.

As with most waterbirds, coots will usually head for the water when approached. However, if they are being pursued on the water, they can be gently encouraged to shore. Once on land, a band of coots can be gently influenced and herded. Too much pressure, though, and they will fly off, possibly out of the area.

Individuals, pairs, and small groups of coots tend to be very insecure and will do one of two things if they don't fly off. The most frightened individuals will look for a place to hide. Quietly, calmly, stealthily, they will try to find a spot where they feel hidden, even if it is only the front half of their body. Coots being stalked have been known to take refuge in clumps of tall grass, in reeds, under footbridges, and in rock crevices where they sit tight – immobile. Coots will stay frozen, even with the threat of being stepped on, so rescuers must be very careful when searching for them in tall grass and reeds.

If there is nowhere to hide, or a coot becomes panicked, it will start to run, and it will keep running, and running. At this point, with the risk of causing more harm by inducing capture myopathy, the bird must not be chased down.

Coots bait in fairly easily. Urban coots can be lured with grain, chips, or breadcrumbs. They are usually quite tolerant of one another and will gather in large groups to feed.

The other rail species are mostly elusive, sheltering in thick vegetation and their reed bed habitat. Rather than trying to drive one of these secretive birds from cover, rescuers should first try luring the bird out at night using audio recordings and spotlighting. Walk-in traps have also proven successful in capturing rails.

Suggested methods for capturing coots: baiting (American coot); walk-in trap; large funnel or drive trap (American coot); land seine (American coot); spotlighting.

Brown pelican (*Pelecanus occidentalis*)

Brown pelicans are plunge divers. They spot prey while flying and plunge into the water to capture a pouch-full of fish. Urbanized pelicans can be found loafing on breakwaters, looking for handouts on fishing piers, near fish cleaning stations, and on bait barges. Brown pelicans that are accustomed to humans can be lured using fish – so close, they can be grabbed by hand.

When catching a pelican by hand, it is not necessary to grab the whole body, just the bill, initially. This capture technique should be considered first, before trying to use a net. Pelicans tend to shy away from nets.

Techniques to entice a pelican close will vary depending on the individuals and how close they are willing to get. Some will walk straight up at the sight of a free meal. Others may take some convincing. Either way, capturers should take a low profile, perhaps on bent knees, with a good amount of baitfish at their disposal. With a target bird in sight and showing some interest a rescuer might want to try looking down, avoiding direct eye contact with the bird, appearing preoccupied with handling a collection of baitfish, perhaps even tossing a couple out a few feet (a meter) or so. If the bird is skittish, it must be allowed time to settle in, receiving a fish from time to time, building the bird's confidence. The rescuer should keep "coiled" until the pelican's bill is within reach. If done smoothly, it is possible to shuttle one bird out of the crowd without alarming the others.

Once the pelican's bill has been grabbed, the bird can be maneuvered so that the rescuer can fold both wings into normal position (Fig. 42). Using their body or leg to keep the closest wing tucked, the capturer can reach over the bird's

Fig. 42 Initially, a pelican can be grabbed by its bill before its wings are restrained.

Fig. 43 A pelican can be carried at the hip, making sure to hold its bill slightly ajar so that it can breathe.

body to gain control of the other. The bird's body can be scooted up under the rescuer's armpit and carried (Fig. 43). With one hand holding the bill, rescuers must make sure to place a finger in between the upper and lower bill so that the bird can breathe.

Suggested methods for capturing pelicans: baiting; land seine; leg snare pole.

Cormorants (Phalacrocoracidae)

Cormorants and shags are fish-eating coastal birds. They can be found in saltwater and freshwater ecosystems. These birds are very skittish and difficult to approach during the day. If pursued too heavily they will disperse from an area.

Urbanized cormornts – those that are used to people, such as the ones that frequent parks or fishing piers – can be lured using live fish, such as goldfish or minnows, in a see through pan. At night, cormorants have been successfully captured by spotlighting, but this must be done carefully as the birds will often head straight for the light.

Suggested methods for capturing cormorants: spotlighting; land seine; net launcher (on land).

Waders

Herons, egrets, and the like are skittish and therefore, if flighted, they are difficult to approach. Spotlighting has not proved successful. However, these species can be lured into a trap zone using bait, such as live goldfish in a shallow pan, or live rodents.

Suggested methods for capturing wading birds: modified bal-chatri; noose mats; single snare; net gun.

Alcids

Alcids, including auklets and murres, are marine, wing-propelled pursuit divers, using their wings to "fly" through the water after fish. While murres and murrelets are more suited for water, their relatives, the auklets, puffins, and guillemots, are able to walk and hop on land.

If found on shore, these birds will try to escape by racing to the water. However, common murres have been successfully stalked, even with the rescuer and net in plain view. They can also be successfully netted from a boat during the day and using spotlighting from a boat at night.

Suggested methods for capturing alcids: spotlighting; stalking; hand netting from a boat.

Birds trapped in structures

Rescuing a bird that is trapped inside a large structure can be quite a challenge. First, the species must be identified. This will give rescuers ideas on how it might be lured down. Second, knowing the length of time the bird has been imprisoned will give rescuers a better idea of its current condition and immediate needs.

During daylight hours, one of the first strategies is to darken the interior of the structure – turning out lights, covering skylights, and, most importantly, darkening windows. The only light entering the building should be from open windows and open doors. As the interior is dimmed, the bird will fly toward the light. However, if not all windowpanes can be blocked or darkened, rescuers might want to consider a more complex strategy, as birds can be seriously injured or killed trying to escape through what they perceive to be an opening to the outside. Sheets of material, such as garden netting, tarpaulin, or bed linens, can be draped in front of closed windows to keep the bird from striking them.

Another thing rescuers should consider is reducing the area the bird has to fly in, quartering off sections using walls of netting or sheets of material. In doing so, at least one rescuer should be ready with a handheld net; birds that fly into material, even netting, have a tendency to drop. Tall barriers can be made out of sections of PVC pipe. The netting can be attached using cable ties. Another option to consider is using heavy-duty helium balloons to raise panels of lightweight garden netting (Fig. 44). The balloons can be anchored in place using bricks.

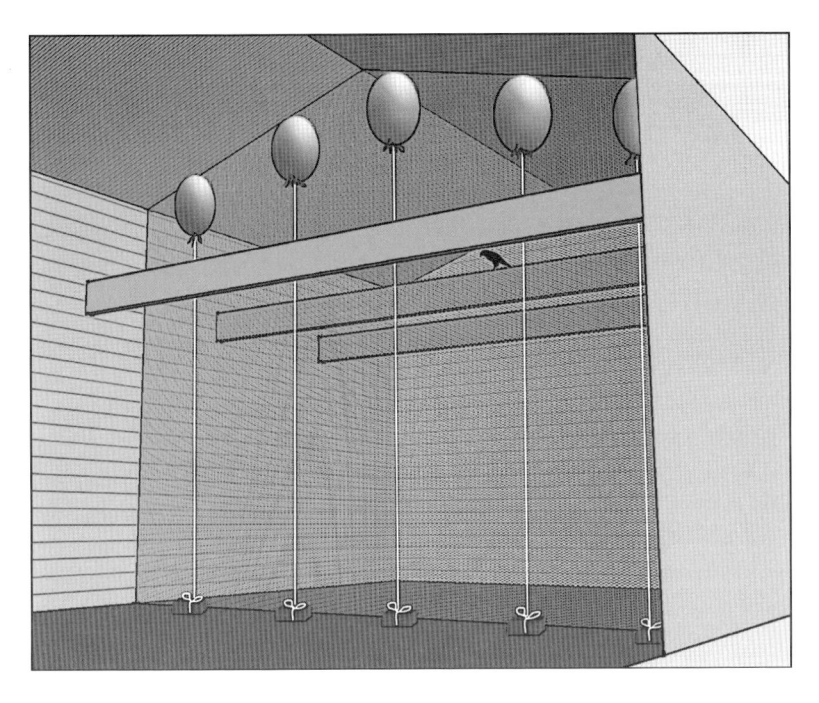

Fig. 44 Helium balloons can be used to raise a barrier of lightweight netting.

When rescuers are trying to get a bird to exit a building on its own, the space must be quiet, with little to no human activity. The bird will not fly down toward open windows or doors with people nearby. To encourage the bird to leave its perch, bait can be placed on the ground near an exit. Helium balloons can also be used to keep a bird from a particular area.

If a daylight mission would pose too great a risk to the bird, rescuers should consider waiting until nightfall when they can try using high-powered flashlights to sneak up on the bird. To reduce risk of injury to the bird, walls of material should be set up ahead of time to prevent the bird from hitting hard surfaces straight on.

Another strategy is to use traps. A small walk-in type of trap or a drop trap is suitable for passerines; bal-chatri or Bartos traps are suited to raptors.

Following study of the bird's behavior, nooses can be set where the bird typically lands or where it habitually roosts at night. In preparation for the bird being snared, rescuers must be ready and able to reach the bird immediately.

Hummingbirds in skylights

If a hummingbird enters a structure with a skylight it will probably fly up toward the window, to what it perceives to be the sky. Unless the skylight is darkened from the outside or the bird is otherwise encouraged away from the skylight, it will keep trying to get through until it becomes too exhausted and near dead. A hummingbird in a skylight is a true emergency requiring immediate action.

If there is no way to darken the skylight, and no way to reach the bird with a long-handled net, rescuers can try sending a helium balloon up to the skylight to get the tiny bird to move away from it. In case that does works, windows and doors need to be open so the bird does not crash headlong into a glass pane.

Window strikes

During the day, darkened windows reflect the outdoors. At night, in a brightly lit city, the same phenomenon occurs. The image appears so lifelike, birds fly straight into windows, often at full speed. Collisions are a leading cause of bird mortality in North America, killing an estimated one billion wild birds annually.

If a window strike doesn't kill a bird instantly, birds can be found unconscious, disoriented, often severely injured and in need of immediate attention. Small, mildly disoriented birds may simply need a quiet spot to recuperate. These casualties can be placed into a dark ventilated container, lined with a non-frayed towel, and left undisturbed for about thirty minutes. If they still appear awkward after resting, they may need veterinary attention and should be transferred without delay.

Property owners can take steps to prevent window strikes. One way is to make the glass pane more visible to birds. Unlike humans, birds are able to see ultraviolet light (UV). UV window decals have proven to reduce collisions. They appear clear to humans, but are visible to birds. They are static-adhering and are commercially available in a variety of shapes and sizes. Another option is to coat windows with a very thin film of mineral or vegetable oil. Oil has a strong UV reflectance. If treating the glass is out of the question, shiny objects such as aluminum pie tins, old compact discs, or strips of Mylar can be suspended outside the window to ward off birds. Another concept, used to keep flying insects away, is a plastic bag filled with water and tied just above a window. The refraction of light may be enough to keep them from hitting the glass.

Ducklings in a pool

Finding a hen and ducklings residing near manmade water features is very common in spring and summer months. Sometimes, it is a situation where the ducklings have been able to get into the water, but not out. Where they are welcomed residents, a rescue may be as simple as giving the ducklings a floating ramp that

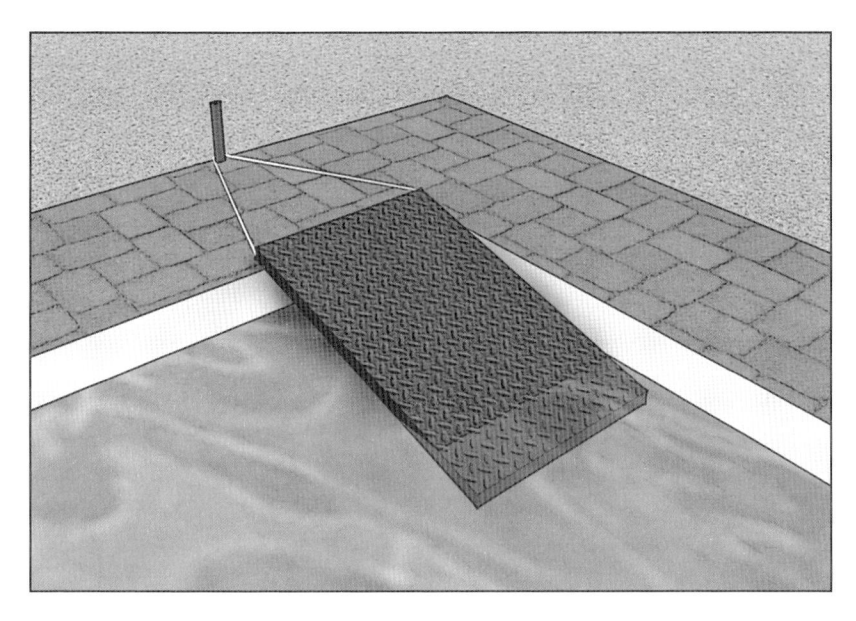

Fig. 45 Ducklings must be provided with a way of getting out of a pool. A doormat or panel of anti-fatigue flooring can be draped over the edge and secured.

allows them to get in and out as they please. This type of ramp can be made from a flexible piece of material, such as an anti-fatigue floor mat. The material should be secured at the pool's edge and allowed to drape gently into the water with at least a foot-long (30 cm) section floating at the surface (Fig. 45). Empty plastic bottles can be secured to it for added buoyancy if necessary.

When the duck family is unwelcome, rescuers will have one or two options available, depending on regulations governing relocation of waterfowl.

For ducks in a pool, the answer might be to escort the family from the pool area to another part of the property and block them from returning to the water. If the ducklings cannot be coaxed out of the pool, they will need to be scooped from the water and then gently herded.

To scoop the ducklings from the water, rescuers will want to use a long-handled hoop net with very soft fine mesh. Using gentle movements the ducklings can be encouraged into a corner. The netter will then scoop under them, under the water, and lift up. Overhead scooping does not work well as even day-old ducklings can dive.

Once out of the water the ducklings and hen can be gently walked out of the backyard using large towels or large pieces of cardboard to herd them. This requires planning and good communication between the rescuers. If necessary, the ducklings can be momentarily confined in a clean plastic bin, open box, or pet carrier, as long as the hen can see and hear them. Typically, the hen will be hard pressed to follow her babies as they are walked a short distance and freed.

Sometimes, the hen must be captured. This is not an easy task. The safest method for capturing her is with a drop trap. Ideally the trap should be introduced into her environment a few days before it is to be used allowing her to become accustomed to it. It should be placed in the area she tends to frequent or where she receives food. When the time comes to set the trap, grain or crumbs can be used to draw her underneath.

Ducks are strong flyers with powerful wings, so the trap might need to be weighted down slightly. A frame of rebar can be attached to the base of the drop trap to help hold it in place against her struggles.

If possible, the ducklings should be contained before the trap is set. There is too great a risk of them being injured by the heavy edge of the trap or by the hen during the capture. When trapped, the hen's attempts to escape may be violent and dangerous to any ducklings in the trap with her. Once contained, the ducklings can be used to coax the hen under the trap.

The ducklings can be placed into a small pet carrier and set just behind the trap, as it is set. The sides of the trap should be blocked, leaving only one way for the hen to get next to her brood – by walking underneath and to the rear.

Once trapped, the hen must be removed quickly and with tremendous care, as she will be panicked. Ideally, the box trap will have an access door on top. Using a bed sheet to cover the entire cage, rescuers will reach in and remove her by hand.

If there is no door, an extra large bed sheet – big enough to envelop the box trap, can be scooted completely under the trap and then, with the help of one or two additional rescuers, the cage can be overturned slowly, keeping the fabric taut at all times. This should result in an upturned trap covered with a tightly drawn bed sheet with the outer edges of the sheet concealing all sides. The hen will be explosive, flying up and hitting the sheet, which must be held firmly against the trap frame at all times.

To retrieve the hen, rescuers will work the fabric down into the trap, covering the hen and immobilizing her. Once secured, she can be lifted up and placed into a separate pet carrier for transport. A hen must never be transported in the same carrier as the ducklings; there is too great a risk of her accidentally injuring them. However, she should be able to see and hear them until they are reunited.

When releasing a hen with her ducklings, the ducklings must be released first as the hen watches. Once all her babies are out of their cage, the hen can be released. Done this way, there is less chance of a mother duck abandoning her brood.

Anecdote: Ducklings and the infinity pool

In keeping with the Migratory Bird Treaty Act, I have always made it a policy to leave mother ducks and ducklings where they are, for the most part. We help residents get babies out of their pools and herded away, but, more often than not, we leave them be, unless they are in imminent danger, or pose a threat to human safety.

In recent years we have worked closely with the US Fish and Wildlife Service and California Department of Fish and Game to come up with a formal response policy for our organization, WildRescue. It is as follows: we will only attempt to capture and relocate wild mallards and their ducklings when the hen and ducklings are in imminent danger of being killed or injured, or when their presence or movement places humans at risk: for example, if they are traveling down a busy road and cannot be herded to a safe location. We may also intervene when we suspect people are going to take matters into their own hands and orphan the ducklings. If and when we relocate, it is within one-half mile (0.8 km) unless we receive explicit permission from the agencies to go beyond that distance.

One spring day in Malibu, California, we received a call from a resident who lived in the hills above Trancus Beach. A hen and ducklings had recently shown up and taken a liking to their beautiful infinity pool. The problem was, the ducklings kept swimming over to the "infinity" edge and toppling into the shallow basin four feet (1.2 m) below, where they were permanently trapped. There was really no way to block the ducks from entering the estate, and the surrounding sagebrush hillsides were inhospitable. We believed the best option was to relocate the family to a nearby lagoon.

When we arrived, the mother duck had taken time off from her maternal duties and was plucking snails from a tidy row of flowering agapanthus. Her eleven ducklings were huddled together in the pool's lower basin, where the water from the infinity pool overflowed. It resembles a moat – long, but very narrow – about two feet (60 cm).

There would be no way of swooping down on the mother duck using a hoop net – the rim would catch on the sides, leaving room for her to scoot from under it. Besides, the net itself poses a great threat – it applies a significant amount of "pressure." Just the sight of it could scare her off for a while.

As we were looking at the pool and developing a plan, the hen flew into the channel to protect her babies from the intruders. We thought to try a unique approach. I decided to use her protective instincts to draw her close – near enough to grab.

The idea was to reduce the threat I represented as a tall, two-legged beast, and pose as a far less threatening creature, something she could possibly ward off.

On my hands and knees I went, managing through tangles of non-native groundcover. I skulked up to the shallow cement channel, near to where the hen was poised with her babies. On eyeing me she bristled, wings slightly out, appearing quite buff. Now, all I had to do was get her to believe she stood a chance of scaring me off.

It worked. Before I was coiled and ready to strike, I was slapped in the face with wet mallard wing – an affront deserving of a reward. What a brave mother. I had to honor her might by backing down and away for a few moments. I had a role to play out.

Approaching the moat again, I led more with my legs, my upper body coiled. The hen, too, was ready – facing me head-on, flared wings. Positioned and ready I recoiled and lunged with speed, grabbing the hen's midsection before she knew what had happened. Folding her beating wings, I brought the matron to my chest, her heart racing. We quickly boxed her up in an awaiting cat carrier.

The ducklings were gathered in a corner of the moat, backpedaling nowhere. They were quickly scooped up with a koi net – a small net of extremely soft fabric. They, too, were placed into a small pet carrier.

The two carriers were loaded into the vehicle and set facing each other. We drove the family to a sheltered creek that fed into a large lagoon. With the mother duck watching, her babies were set free at the edge of the water. They sprang and tumbled and flopped out of the carrier and into the placid water. Seconds later, the door to the hen's cage was opened. She hesitated, and then in a burst she shot out, and straight to the water where her babies were. We watched, smiling, as the family of twelve headed toward the far bank of willows and wild watercress.

Birds entangled in fishing tackle

As a rule, unless a bird is freshly hooked or snagged in fishing gear, it should receive a complete examination and assessment. Quite often, victims of entanglements are underweight and suffering from infection. Once untangled, though, if a bird presents with no major injuries or signs of illness and is in excellent body condition, its immediate release can be considered. A licensed wildlife rehabilitator or wildlife veterinarian should make the final decision.

If the bird is severely tangled and in poor shape, cold and near shock, the most that the rescuers should do is snip any line that is causing the animal immediate pain or discomfort, leaving the rest for a later time when the bird is in a stable condition.

When a bird is able to endure "man-handling" to be freed of line and hooks, rescuers will want to examine the bird carefully and methodically snip and remove line. For hooks that are deeply imbedded or otherwise unable to be removed at the time, rescuers will want to leave a tracer of fishing line so it can be located again easily.

Rodenticide poisoning

In urban and agricultural areas, birds that consume rodents are at risk of secondary poisoning from the widespread use of powerful rodenticides. The majority of rodenticides used in the United States contain anti-coagulants like Brodifacoum, Bromadiolone, and Difethialone that cause internal bleeding and death. These are second-generation anticoagulant rodenticides (SGAR) that were first introduced in the 1970s after it was discovered that rodents had become somewhat resistant to the first-generation poisons, such as warfarin.

Although these new and improved rodenticides are supposed to deliver a lethal dose after one meal, poisoned rodents are known to spend time in open areas before dying. A staggering rodent in an open field can be irresistible to a hungry predator.

After consuming a poison-laden meal, a predator may not exhibit signs of secondary poisoning right away. However, these chemicals can and do accumulate in the liver, at sublethal levels, and can persist in an animal's system for weeks. During that time, if a predator ingests multiple doses, it can cause the animal to hemorrhage and die.

Even sublethal levels, over time, can interfere with an animal's ability to survive. Research has shown that sublethal, chronic exposure to these anticoagulant poisons can compromise an animal's immune system, increasing its susceptibility to disease.

For wildlife rescuers it can be difficult to determine if a bird is suffering from rodenticide poisoning until it is tested. Nevertheless, signs of weakness, such as uncharacteristically low or short-distance flight, can be indicative of a problem. In urban or agricultural areas, rodenticide poisoning should be considered in the diagnosis.

Shot through with a projectile

A mobile bird that has a projectile, such as an arrow, protruding from its body presents a unique challenge. The use of a net or netting material that could snag on the foreign object should be avoided. In these situations, then, the best option is to use a large wood or wire drop trap or large walk-in trap – large enough that the projectile will not get caught up. The idea is to get the bird into a confined space where it can then be captured by hand or using a sheet. A snare has also proved a successful method of capture in these delicate situations.

Anecdote: The story of Pinky the turkey

It was near Thanksgiving when some residents of a developed ridgeline in Castro Valley, CA, noticed that one of their beloved wild turkeys had been shot through with an arrow. The sleek carbon fiber hunting arrow had pierced through the bird's body and was sticking out its back. Even so, the bird was still able to get around, limping, and he was still able to fly.

"Pinky," as he is known by his admirers, roamed free with a large rafter of other wild turkeys. Moving through the oak woodland like a herd of feathered dinosaurs, the turkeys visited the hilltop community regularly, checking out a few particularly inviting backyards for scattered birdseed and special handouts.

A middle-aged couple, whose property is visited daily, knew this Pinky well, since he and his brother were poults. When they found the bird injured and suffering, they were quick to reach out for help. Fortunately, as you'll see, they couldn't find anyone willing to try to rescue the bird, being told by "professionals" that if a bird can still fly, it cannot be helped. After numerous disheartening calls, the couple gave up. All they could do was watch as the great bird tried to keep up.

Weeks later, through forwarded e-mails, word of Pinky and his dilemma reached WildRescue. Duane and I were quick to take on the challenge. With the arrow sticking out of the body, though, netting the bird was not an option. The arrow would become entangled in the mesh. We were also dealing with a very strong and very flighty game bird. The property was also a two-hour drive from home.

Our first course of action was to gather information about the location and the surroundings. Through e-mail we were sent pictures that gave us a rough idea, enough to start building a plan. The next step was to visit the site and get a real feel for how the capture might play out. We also wanted to get a look at Pinky.

Sure enough, during our visit we got a glimpse of Pinky and his family. They crossed the dappled, leaf-covered ground some 50 yards (45 m) from the property in silence, finally climbing from the dell onto a 4 feet (1.2 m) wide shelf – the retaining wall for the backyard. The yard itself was bordered by a wrought iron fence, with no gate. To get into the yard, the birds had to fly onto the railing and then down – something Pinky was not about to do.

Having had a good look at the layout, Duane and I dropped about $100.00 on material at the nearby hardware store and got to work on building a trap. It was well into December and very chilly.

Using the wrought iron fence for one side, we erected a "run," or pen, made of 4 feet (1.2 m) tall green garden fencing secured to metal stakes. The top was covered with chicken wire – garden netting is too flimsy to hold these powerful birds. Duane rigged a bungee cord-loaded door that would slam shut and lock with just a pull of a string. Seed was placed deep into the long, nearly 14 feet (4.25 m), enclosure, at a location where the birds are used to feeding.

It did not take many days for the turkeys to become accustomed to the cage. They went inside and down its length without hesitation. Before long, Pinky, too, would enter. One afternoon the homeowner found Pinky inside, along with one other male. He triggered the door, trapping them both. Sheets and blankets were used to cover the enclosure in the hope of reducing their stress.

By the time Duane and I arrived, it was dark. We prepared the bird's carrier, a large plastic dog crate, and using a couple of ladders we carried it over the tall wrought iron fence to the ledge. A recent rain shower made the ledge quite slick.

With a flashlight and a couple of sheets in hand, Duane entered the pen and allowed the large, non-injured turkey to slip out the open door. We then carried the dog crate inside and locked the gate behind us. Using a sheet stretched between us, Duane and I tried to corner the frightened animal, slowly moving in on him until we could cover him with the sheet. It took a few attempts with him scooting past us before we were finally able to grab hold of him. He was extremely strong, arching up with his powerful legs, and feathers started flying everywhere. As a defense mechanism, some species of birds will readily "drop" their feathers when attacked.

Within a couple of minutes Duane had cut through the arrow so that Pinky would fit into the dog crate. It took the two of us to lift him in. With the couple's help we managed the "live" crate over the fence, and into our rescue van. Even though it was bitterly cold outside, we kept the car cool for his hour-long journey to International Bird Rescue in Fairfield, California.

The next morning, with puffs of feathers all around, Pinky was sedated and examined. The arrow had entered his back and exited his lower chest area.

There was little bleeding – the body had begun encapsulating the foreign object. X-rays, however, revealed that the impact of the arrow had fractured his femur. Miraculously, though, due to the delay in his rescue, his bones had mended. The arrow was removed and the canal debrided and flushed, and Pinky was brought out of anesthesia.

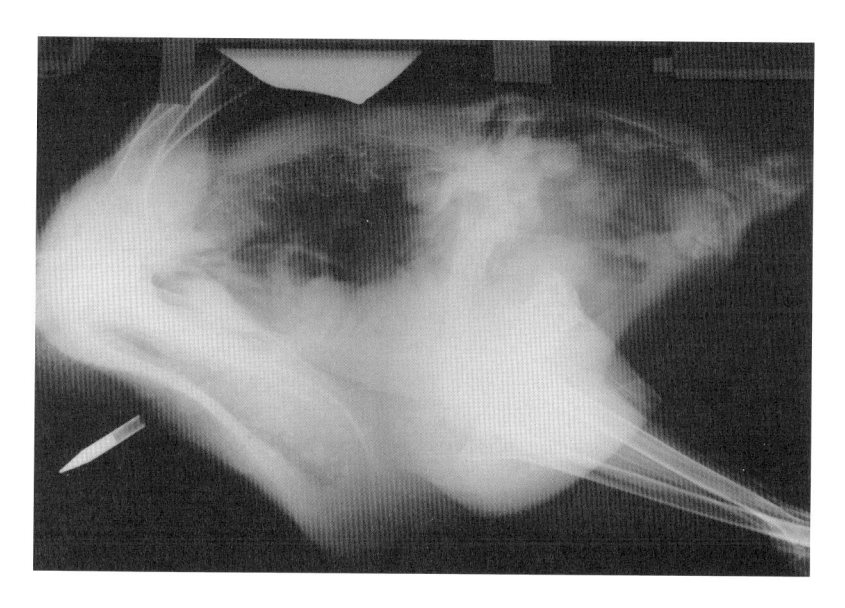

Turkeys are ballistic. They are big and powerful game birds that react explosively to perceived danger. Like deer, turkeys can easily sustain fatal self-inflicted injuries in captivity, so they require special housing. Even so, the risks of a captive setting were higher than the ones Pinky would risk in the wild.

The following afternoon, just a couple of days before Christmas, Pinky was released within 20 yards (18 m) of where he was captured. The second the door to the kennel was opened Pinky bounded from the box and straight into the air, landing high in a eucalyptus tree.

Pinky has been seen on numerous occasions and has all but lost his limp. He has developed into a beautiful bronze gobbler with a long black "beard" and flashy iridescent feathers.

Glue traps

There are various types of commercial sticky traps used to control pests. Some are meant to catch flying insects while others are used to snag small rodents. Every once in a while a bird will become stuck on one of these devices.

The first course of action is to try to relieve the bird of serious pain and discomfort. Ideally, it should be removed from the contraption and helped to regain as close to normal body posture and movement as possible. However, it may be best to leave this extremely delicate task to a professional wildlife rehabilitator. Birds in these situations are under a tremendous amount of stress and may be close to shock. If the bird is weak and cannot withstand the process of being removed from the trap, rescuers might consider first aid, possibly a few drops of hydrating solution and a rest from the sights and sounds of humans.

Removing birds from sticky traps requires a great deal of skill and patience. First, the sticky parts of the trap should be doused with a generous amount of cornmeal. This will prevent the bird from getting stuck any further. With one person gently restraining the bird, another can work on freeing its feathers and body parts, using cornmeal to prevent them from becoming re-glued. This must be done extremely carefully as it is easy to damage ligaments, break bones, or tear a bird's paper-thin skin.

Small amounts of a product called methyl soyate, a relatively safe soy-based solvent, can be used at the contact points to help break down the glue. This product is often used as a pre-treatment for oil-soaked birds. It can be warmed slightly, and a very little amount can be applied using a Q-tip. Sticky Out, by Elmers, and canola oil have also proven helpful in releasing birds from the glue traps. As a precaution, a veterinarian or toxicologist should be consulted before any product is used that will remain on the bird for an extended period of time.

Once the bird has been freed from the trap it can be lightly dusted with cornmeal if needed. This might help it gain normal posture.

Avian botulism

Avian botulism is a paralytic disease of wild birds caused by ingestion of a toxin produced by *Clostridium botulinum*, an oxygen-intolerant bacterium that persists in the environment as dormant spores until conditions are just right. Over the past 35 years, avian botulism outbreaks in the United States have increased significantly. Worldwide, avian botulism is a significant disease of migratory birds.

Most botulism outbreaks occur in summer and fall, when warm weather and abundant organic matter can create a perfect environment suitable for germination; the toxin is produced once the spores begin to develop. A rotting carcass offers a prime start for the bacteria to multiply and enter the food chain. Invertebrates, like maggots, feed on the carcass and act to concentrate the toxin. Waterfowl and other species then feed on the invertebrates.

Avian botulism causes paralysis of the muscular and nervous systems. Birds can lose control of their legs, wings, eyelids, and neck. Critically ill birds will lie prostrate with little eyelid response. Less compromised individuals will have difficulty walking and holding their head up, and their eyelids will be slow to respond.

Because they are unable to escape the elements and cool themselves, many birds die from overheating. When responding to an avian botulism outbreak, search and rescue efforts must be conducted throughout the day, beginning early in the morning. Areas where birds are likely to be found should be checked every three to four hours.

Treatment for avian botulism is aggressive fluid therapy. Sterile saline is used to flush the eyes of birds that have lost the use of their eyelids. Ophthalmic lubricants are administered until they regain use of their eyes.

Lead poisoning

Lead toxicosis is common in waterfowl, raptors, and some seabirds. Typically, birds are exposed to lead through ingestion of spent lead bullets or pellets and lead fishing sinkers. Once the level of lead absorbed into its system reaches a toxic level, a bird will begin to exhibit symptoms, which can include weakness and an inability or reluctance to fly. Early on, a bird suffering from lead poisoning may hold both of its wings oddly, in what can be described as a "roof-shaped" position. In later stages its wings will droop. Bright green, bile-stained feces are also suggestive of lead poisoning.

Domoic acid poisoning (DAP)

Domoic acid (DA) is a neurotoxin produced by a few species in the genus of single-celled phytoplankton, or diatom, called *Pseudo-nitzschia*. The ubiquitous *Pseudo-nitzschia* are found throughout the world and, like other plankton species, they help fuel the marine food web. Although microscopic in size, diatoms are extremely important to life on earth – during photosynthesis, they take in carbon dioxide and generate oxygen.

When conditions are just so, with the right amount of sunshine and nutrients in the water, the phytoplankton will "bloom." In huge numbers they can discolor the water, which is sometimes referred to as "red tide." When the bloom consists of a species that can be potentially damaging to humans or the environment, they are referred to as harmful algal blooms, or HABs. Although algal blooms are a natural phenomenon, some scientists cite pollution, the destruction of wetlands, and global warming as contributing to the notable increase in HABs.

When a domoic acid-producing species of *Pseudo-nitzschia* blooms, it may or may not produce DA; researchers are still trying to figure out exactly what activates its production. What they have concluded through recent research is that urea and iron can stimulate its growth. When a bloom does consists of diatoms that are producing DA in high concentrations, the toxin can build up in the food chain, posing a threat to humans and marine animals. Planktivorous fish, such as anchovies, are thought to be vectors, passing on high levels of DA to predators such as pelicans and sea lions.

Brown pelicans suffering from domoic acid toxicosis will usually, but not always, display neurological disorders including a characteristic side-to-side weaving of the head, scratching, wryneck, clenched toes, and a slumped posture. They also tend to experience disorientation and issues with mobility and navigation. During DA events, pelicans have been found walking in city streets or on rooftops and backyards many miles inland. In severe cases, they have been known to literally fall out of the sky. Nevertheless, many remain flighted and able to elude capture.

Other marine birds, including cormorants, loons, and grebes, have tested positive for DAP without showing symptoms other than the fact that they were easily captured.

In the early stages of a possible domoic acid event, when reports start coming in about pelicans behaving oddly or sea lions seizing, rescue organizations should start networking together and sharing information about sightings. This will help build a thorough proactive search and rescue plan. Where possible, teams of rescuers should check "hot spots," or loafing sites, for disabled birds. If ill marine birds can be found and captured, treatment is often successful.

With marine birds, aggressive fluid therapy can flush the toxin from the animal's system. Once an animal is exhibiting symptoms, though, there is no time for delay. The longer the toxin is in the body, the more chance there is of it causing severe and permanent damage to the brain. This is especially the case with marine mammals.

Search and rescue organizations will want to work with the rehabilitation center that will be receiving the patients and follow their prescribed treatment. They may recommend starting fluid therapy immediately, prior to transfer, or prefer for the animals to wait until they arrive.

The first confirmed DA mortality event occurred in Monterey, California, in 1991, when brown pelicans and Brandt's cormorants began displaying odd behavior. Since then, domoic acid events along the California coast have become almost seasonal. Globally, HABs are increasing in frequency, magnitude, and duration. Unfortunately, because they have yet to be directly and clearly attributable to Man, they are considered "natural" events, and no government agency is accepting responsibility for the response to wildlife casualties when they occur. Once again, the non-government organizations are placed in the position of having to respond with no outside support or resources. For large-scale events, the challenge can be overwhelming.

"Sea slime"

In recent years, another type of microplankton, a dinoflagellate, has been the cause of at least two widespread seabird mortality events along the Pacific Coast of North America. The species responsible is a mushroom-shaped single-celled organism called *Akashiwo sanguinea*. It usually blooms without incident. However, when conditions are just right, it can make the sea deadly for birds.

The first documented case occurred in November 2007, in Monterey Bay, California, after one of the first heavy rainfalls of the season, leading researchers to believe that *A. sanguinea* is stimulated by low salinity and a certain nutrient load. The second incident occurred in October 2009, off the coast of Oregon and Washington. Both incidents had weather in common – the seas were turbulent from recent storms. High winds and choppy seas are thought to have helped break apart the phytoplankton, whipping the organic soup into a surfactant-like foam. Birds that came into contact with the slimy water lost their ability to stay waterproof and struggled to keep afloat. Thousands of birds were found dead and dying along miles of coastline.

Of the hundreds that were recovered alive, many were severely hypothermic from being wet, and hypoglycemic. Some exhibited a slimy pale yellow-green substance on their feathers, mostly on the breast. Given time, the substance would dry, leaving a pale, crusty residue on the feathers. The proteinaceous material was found to be relatively non-toxic and easy to remove using the washing procedures normally used in treating oiled birds.

Oil and petroleum products

Feathers act like shingles on a roof. They overlap like shingles. It is their structure and alignment that keeps a bird sheltered from the elements and allows it to effectively thermoregulate. Certain substances, such as soap, oil, or even sugar water, can cause feathers to, essentially, collapse or stick together, creating an opening. Through this gap in a bird's protective layer, precious body heat escapes and water and air can reach the bird's skin. Unable to keep warm, the bird's body starts burning fuel, using up fat reserves and muscle to keep warm. Compounding the problem, an oiled bird will spend extra energy trying to right its plumage, spending an inordinate amount of time preening rather than feeding.

The degree to which oiling will impact a bird is influenced by these factors: the toxicity of the material, the degree of coverage, the species, the age of the animal, the climate, how quickly a bird is recovered from the field, and the quality of rehabilitative care.

Given these variables it is often difficult for rescuers to predict overall survivability. It is easier to evaluate impact and survivability in three phases: immediate – after exposure and before the bird is rescued; survival through the rehabilitation process; and long-term survival – after the bird is released.

The more toxic the product, the greater the potential for medical complications, such as respiratory issues, gastrointestinal issues, or the sloughing of "burned" skin. This means a longer convalescence and a prolonged stay in captivity. Certain species fare better in a captive setting than others. Given adequate resources for their care, pelicans, gulls, penguins, alcids, and most waterfowl can fare well.

Toxicity can also impact long-term survival. Extremely hazardous substances, such as dielectric fluids from transformers, which contain polychlorinated biphenyls (PCBs), are known to cause a variety of reproductive problems in birds. However, even a patch of an innocuous substance can be fatal. It all depends on the other variables.

The degree of oiling – what percentage of a bird's body is covered with a substance, is also a major factor. A bird that is moderately covered in, for example, vegetable oil, yet able to elude capture has less chance of surviving than the same bird caked in crude oil. This has to do with how long they remain in the environment, vulnerable to the elements. In general, the most heavily coated birds often stand a better chance because they can be collected quickly.

The amount of time an oiled bird can survive in the environment will depend greatly on the climate. In cold weather, a dime-sized spot of oil can be fatal. In warm climates, modestly oiled birds can sometimes function normally for a number of days, expanding what is referred to as the "window of opportunity" – the amount of time rescuers have to recover oiled birds from the environment before they succumb.

Even an oiled bird that can fly and is observed feeding will eventually become debilitated if it cannot thermoregulate effectively. The more debilitated a bird is by the time it is admitted for rehabilitation, the more difficult its recovery. Therefore, birds with any degree of oiling that compromises their ability to thermoregulate must be rescued, sooner rather than later.

Sometimes, usually after rain, birds can be found "waterlogged," with their feathers separated, clumped, and matted but not discolored; this can be caused by a thin film of oil, or sheen. Depending on the substance, birds can dry out within an hour or so and appear fairly normal although their feathers may look "disrupted." During the Deepwater Horizon oil spill in the Gulf of Mexico, birds were found in this condition – wet to the skin. It was speculated but never proven that this was because of Corexit, an oil spill dispersant. Again, any bird that appears wet to the skin, even in patches, should be collected if its ability to thermoregulate is impaired.

Occasionally, a bird's feathers will appear lightly smudged with oil but not deeply matted or clumped. This can be described as surface oiling. In very warm climates, some birds with surface oiling may be able to survive if not captured. The oil may fade in color, but will not completely disappear until the bird molts the contaminated feathers.

When judging how the oil and exposure to the elements will impact a bird, its age must also be considered. In general, hatchling chicks have a reduced thermoregulatory capacity. Contaminated chicks may also be vulnerable to overheating and sunburn. As a general rule, collection is advised when oiling is greater than 15% of the chick's body. One thing for rescuers to be aware of: young colonial-nesters may be found caked with excrement, which can resemble oiling, but it will be crusty and washes off easily.

As for fledglings, they have a small window of time in which to learn to successfully feed. Like the adults, if the degree of oiling causes them to preen rather than forage, they should be collected sooner rather than later. Reduced caloric intake for a young bird can be extremely detrimental.

Most oiled birds will present into rehabilitation in a weakened state. This is exactly the reason they must not be washed right away. It will be too much for their system and they will die. The washing process is extremely stressful for wild birds, comparable to a human receiving a bubble bath from a grizzly bear. Decades of oiled wildlife rehabilitation studies have proven that allowing birds a few days to build up strength results in a higher survival rate.

This initial care does not have to wait until the bird is admitted into a definitive care facility but can be and should be started as soon as possible, even in the field. Once captured, oiled birds may need immediate attention. If heavily oiled, the nares and mouth may need to be cleared of oil and debris. They can be gently wiped with gauze, a Q-tip, or a rag. If birds are cold, they should be supplied with supplemental heat, and if they are overheated, they should be cooled. Most oiled birds will be dehydrated and may benefit from administration of fluids.

To increase the number of birds that survive, oiled wildlife recovery efforts must be proactive. Instead of waiting for oiled birds to be reported, search and rescue teams must actively patrol for casualties. Recovery teams will use their knowledge of the species to guide them on where and when to look for oiled animals. They will also use information on the oil and its trajectory in setting their search priorities.

Depending on the type of oil, heavily coated birds may be unable to move, much less fly. If they are near shore they will strand on beaches and breakwaters – anywhere they can get out of the water. In some cases, a bird may be so heavily coated with oil that it no longer resembles a bird; small birds that are caked in heavy oil will wash ashore as globs. These casualties are usually found early on during a spill event.

In the open ocean, heavily oiled birds that cannot fly will drown. They will be desperate to haul out of the water on anything, be it a floating piece of debris or even a manned boat. Search and rescue crews should frequently check for disabled birds near heavy oil concentrations, if possible. Oiled birds that are still able to fly will usually travel to where they feel sheltered and safe, away from oil spill cleanup operations. Birds that are able to, will head for quieter, safer locations with fewer humans. Oiled birds have been known to travel over 50 miles (80 km) from where they were first oiled. Search and rescue effort must take dispersal into account when planning for their recovery.

Typically, when conducting large-scale oiled wildlife rescue missions, a region is divided into sections. Search and rescue teams are assigned to a particular area matching their skills and equipment with the terrain and species they expect to encounter.

The advantage of dedicating teams to specific areas day after day is that it allows rescuers to become intimately familiar with the geography and the animals living there. They will be sensitive to trends and changes, and animals will tend to become accustomed to their routine presence in the environment.

As a rule, recovery teams debrief once a day, usually in the evenings. Search and rescue teams share sightings, helping to build the next day's plan. Evening meetings allow search and rescue operations to begin early, at dusk or just before, making the most of optimum conditions. Mornings are often when the weather is calmest, and the birds are usually cold, hungry, and easier to capture.

Successful oiled wildlife recovery efforts are based on sound planning. An oiled wildlife search and rescue plan will account for the type of oil, the species, the climate, and available resources. In general, search and rescue operations should commence immediately, with teams of experts in the field on the first day, even if it is only to assess the situation. The number of rescue personnel should match, if not exceed, the anticipated number needed to adequately survey the area and collect the projected number of casualties. Too often there are delays in getting an adequate number of rescuers into the field. It is far better to allow for an abundance of resources than for too few.

An oiled wildlife search and rescue plan will include a section on safety that will spell out polices and required protective equipment. Within the Safety Plan will be a section on Communications that will detail how units will communicate with one another. There should also be a Reconnaissance Plan that specifies how surveys will be conducted. A Recovery Plan should detail the parameters for the collection of live oiled wildlife as well as oiled carcasses.

The casualties of a pollution spill are essentially evidence of a crime. There are typically chain-of-custody protocols and specific instructions on how to collect and process recovered carcasses. Usually, carcasses are wrapped individually in paper bags and then bagged again inside a second paper bag. Plastic is avoided as it can interact with petroleum, changing the product's hydrocarbon "fingerprint," and ruining the evidence. If sturdy paper bags are not available, the animal's body may be wrapped in aluminum foil before being placed into a plastic bag. Each body bag must be labeled with the date, time, location using GPS coordinates if possible, species (if known), and collector's name.

An additional part of the planning process will involve safety training. In the United States, oiled wildlife responders are considered hazardous waste workers. Because of this, oiled wildlife response personnel will most likely be required to have a specific level of training and possibly 24-hour certification. However, the level of training will depend on the hazards. In some cases, wildlife rescue organizations can provide in-house training that will satisfy regulatory requirements.

Wildlife search and rescue organizations may be able to respond to incidents involving natural petroleum seeps without involvement from outside agencies. However, when a spill of oil or other pollutant occurs it will probably be responded to by numerous agencies through what is called the Incident Command System, or ICS. The ICS is an effective tool for managing small and large emergencies, especially ones that involve multiple jurisdictions. It is a framework within which multiple responding agencies can work together efficiently and effectively. More complex responses may be expanded into a Unified Command.

Very simply, at the top of the pyramid-shaped command structure is the Command or Unified Command. The Operations Section sits below the Incident Command. Wildlife recovery efforts are managed through the "Wildlife Branch" of the Operations Section.

Handling and restraint of wild birds

Unlike mammals, birds do not have a diaphragm muscle, but take breaths by moving their sternum. When handling a bird, care must be taken not to constrict its ability to move its chest to breathe, especially since it will be under stress, with elevated respiration.

Another thing to consider is a bird's feathers. Just a small amount of an oily or soapy substance can disrupt the structure of a feather, allowing cold air and water to seep in next to the bird's body. Therefore, it is imperative that anything that comes into contact with a bird must be free of contaminants. This means that bare hands must be clean, dry, and free of lotion or soap residue.

Fig. 46 How to hold a hummingbird. Thanks to Kane Bridges for permission to use his photographs.

Very small birds, such as hummingbirds, may be grasped like one might pick up a finger sandwich, secured between the thumb and fingers (Fig. 46). They may also be very carefully restrained in a paper tube made from a rolled index card. They must not be restrained this way for more than a few minutes.

Most garden birds may be safely clutched in one hand, delicately but securely. One way to hold a small bird is with its back against the palm of the hand with its head resting gently between the index finger and middle finger (Fig. 47). This hold allows the sternum to move freely.

Another method for holding small birds calls for gripping the bird's upper legs gently between the middle and index finger, supporting the bird in an upright position (Fig. 48). This hold is widely used by bird ringers, or banders. It takes greater skill than the other hold and should not be used on severely injured birds.

A cupping method can be used on small, heavy bodied flightless birds, such as ducklings. The body is held gently, cupped in one palm, while the other hand covers the bird's wings and back, preventing it from jumping.

Another method of restraint that can be used on medium-sized birds and small raptors is sometimes called the "popsicle hold," or "bouquet hold." In this method of restraint the bird is gripped with its primary feathers, tail, and legs held in the palm of one hand (Fig. 49). This is not suitable for heavy-bodied birds.

More robust birds, such as small ducks, can be controlled with two hands much like grasping a football. Care should be taken to keep the bird's wings held in their

Fig. 47 One method for holding a small bird.

Fig. 48 Another method for holding a small bird.

Fig. 49 A method of restraint for medium-sized birds and small raptors.

normal position (Fig. 50). A lightweight towel can be draped over the bird's head and body before it is restrained. This will help shield the bird from visual stimuli and aid the handler in keeping the wings tucked in their normal position. The one thing to watch out for is that the animal does not overheat.

When handling birds, it is usually important to have control of the bird's head as well as its body. When handling mid-size to large birds, this will require both hands – one hand controlling the torso, the other controlling the head. The head may be restrained in a couple of different ways depending on the species. The hand can gently but firmly cup the skull, or a soft grip can be taken around the neck at the base of the skull. The latter may look like a stranglehold, but there should be little or no pressure on the neck.

When handling birds with long pointed bills, such as loons, grebes, or herons, the head can be controlled in what is sometimes referred to as the "cigarette hold," also known as the "martini hold." With the thumb positioned on the back of the bird's head, its bill rests securely between the index finger and middle finger (Fig. 51). This can give the handler good control of the bird's head and bill. This

Fig. 50   A method of restraint for robust birds, such as ducks.

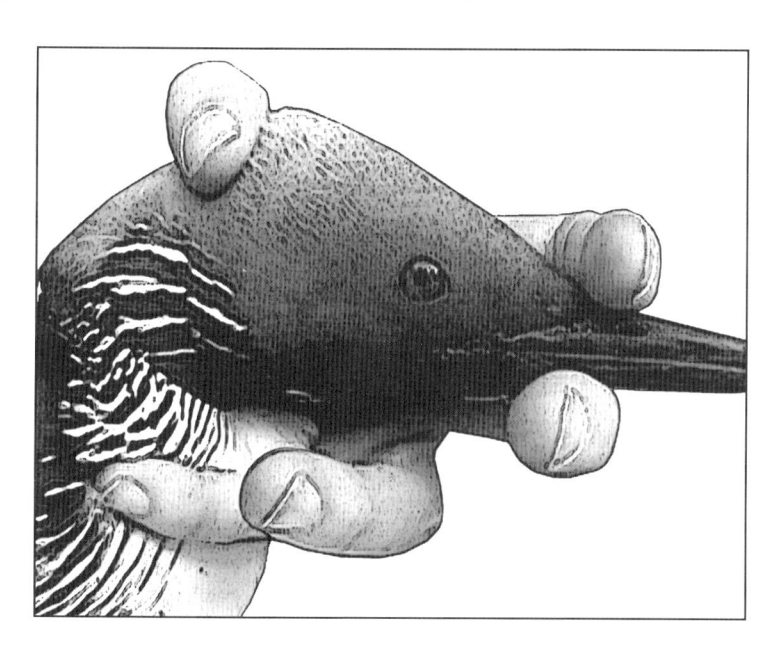

Fig. 51   The martini hold.

hold is not suitable for all types of long-billed birds, however. Birds with small nares, such as cormorants, cannot breathe if held in this fashion. A cormorant must be allowed to take in air through its mouth.

Pelicans, like their relatives the cormorants, require special handling due to their small nares. When restraining a pelican's head, care must be taken to allow the animal to breathe through its mouth. Softly holding the top and bottom bill with one hand, about half way down its length, the handler should have at least one finger keeping the bill from closing completely (see Fig. 43).

When carrying a medium- to large-sized bird, the head may be controlled by one hand with the bird's body cradled against the hip, low enough that its head is out of striking rage of the handler's face.

When handling raptors, the handler should wear protective gloves of the appropriate thickness. If the leather is too thin the talons can pierce through to the skin; too thick and the handler will lose dexterity.

To initially gain control of a grounded raptor, it is often helpful to throw a lightweight towel or sheet over the bird. The fabric will block its sight of the handler and help the bird's wings into normal position. The next step is for the handler to gain control of the bird's legs.

Small raptors can be grasped with both legs held gently in one hand, high up on the legs where they meet the body. The handler must use caution that the legs are held slightly apart, not crossed or rubbing together. This can be accomplished by placing a finger in between (Fig. 52). Once the legs are restrained, a lightweight towel can be draped over the raptor's head to reduce stress. The other hand can keep the wings held in their normal position. Another method of keeping the wings tucked is to position the raptor's back up against the handler's torso. Larger raptors will require the use of two hands, one on each leg.

Another method of restraining raptors or other large birds involves a specially made cloth device called an "abba," also spelled aba. The abba is an ancient method for handling and restraining birds without damaging their plumage. The device is simply made out of a rectangular piece of lightweight cotton fabric. The corners of one end are folded and stitched to form two pockets meant to hold a bird's wrists. In a sense, it is like a straitjacket.

When first applied, the bird will look as though it is wearing a cape. A long strip of elastic fabric, centered on the backside of the cape, is used to secure the abba around the primaries and the bird's upper legs. The abba secures the bird in a popsicle or bouquet hold. The device is only meant as an aid for handlers, not for extended use. Birds being restrained using the abba device should never be left unattended.

Game birds, such as doves, quails and turkeys, can be extremely difficult to handle. Their alarm response can be violent and explosive and they will tend to drop feathers in reaction to being handled. Extreme care must be taken when

Fig. 52 Proper restraint of a raptor.

planning to handle these robust prey species. They are very strong flyers with powerful wings and muscular legs. They can easily injure themselves when trying to evade capture or struggling under restraint. Although use of a sheet or towel may be helpful in handling these species, they will tend to overheat very quickly. Diazepam at 10 mg/kg intramuscularly or intravenously has been used to quieten these birds long enough for an examination.

Coots and their relatives have a tendency to use their legs much like a cat uses its back legs to rake at an attacker's grip. Care must be taken to control a coot's long legs when handled.

Parrots can deliver bone-crushing bites. As with most birds and mammals, the base of the skull and lower mandible can be controlled in a hold. The feet of a parrot pose little threat. Momentarily wrapping a parrot in a heavy towel can help handlers gain an appropriate hold. Giving a parrot something to bite down on, like a flap of leather, may help reduce its struggling during restraint.

Processing from nets and housing

After a bird is secured in a net, a lightweight sheet or towel can be thrown over the bird to reduce visual stressors. To process it from the net, one handler will want to secure the bird's body, keeping the wings folded in their normal position, while the other handler manages under the net to secure the head and legs. Once they have control of the bird, the net can be lifted away. A lightweight towel should be draped over the bird's head to reduce visual stressors.

To safely place a flighted bird into a box or carrier, the door or sides of the box can be held closed against the handler's arms as they are placing it inside. This reduces the chance that it will escape. Once secured in the carrier, it is common practice to scribe on the outside of the box what type of bird it is, when and where it was picked up, and any other relevant details. This helps the next handler to prepare for what is inside.

Every so often a rescuer will retrieve a bird from a carrier or an enclosure. This can be a bit tricky, especially if it is flighted.

Grebes, loons, and the like are relatively easy to catch up. A towel can be flung over them, making sure the majority of the towel lands over their head. The handler will then clutch the head in one hand and scoop up the bird's body with the other. They will adjust to a more proper restraint position once they have the bird out of the carrier.

Removing a flighted bird from a folding pet carrier requires a little finesse. A large sheet or towel can be draped over the exterior of the box before it is opened. As the handler unlatches the box and starts to open the lid, they will want to block the bird's escape, using the towel or sheet. As the lid is opened wider, the fabric should be lowered into and on top of the bird. The handler can then gain control of the bird's wings and its head, adjusting to a more appropriate grip once the bird is removed.

In larger enclosures, especially where there are multiple birds being housed, a large sheet can be used to section off portions to reduce chasing and panicking of all of the birds. Sheets or tarps can also be used to corral birds to one corner where they can be easily picked up by hand.

Short-term and temporary housing for wild birds

This section offers suggestions on how to temporarily confine birds after they are captured. It is divided into two parts. The first describes confinement suitable for the short term only, just long enough to get the bird from a capture site to a vehicle or staging area where it will be transferred into more appropriate temporary housing.

A lightweight pillowcase can be used to hold a single small to medium-sized bird, other than a raptor. The deeper the sock, the better. The open end of the pillowcase gets tied in a knot, leaving enough room for the bird to move. The bundle may be carried at waist-level with a hand cupped underneath, supporting the bird's weight. The bundle can also be placed in a container. However, birds carried in pillowcase sacks must be kept upright and must have ample ventilation – they must never be stacked one on top of another. It is very important that the sack be kept dry.

Paper shopping bags with air holes can be used, short term, to carry small songbirds. A non-frayed, tightly woven, kitchen towel or pillowcase may be placed on the bottom of a bag. The top of the bag can be folded and clamped shut.

Different sizes of fabric sacks or mesh bags, such as mosquito head-nets, dive bags, or laundry sacks, can be used to secure various sizes of birds for a short time. Heavier birds may need to be supported so they don't topple and flail while being carried. With mesh bags, a lightweight towel can be used as a visual barrier when necessary, as long as it does not impair ventilation.

In cool climates, large bodied birds, such as swans and geese, can be temporarily secured in fabric for quick transport out of the field. A bed sheet can be used to lightly wrap a bird's body, fixing its wings but leaving the head, chest, and tail area free. Heavy-duty tape is used to hold down the ends. Canvas bags made for just this purpose are commercially available.

As for temporary housing, as a rule, an animal should be able to maintain normal body posture and have enough room to turn around without hitting the walls. Pelicans, however, should be given enough room to extend their bills. This is especially important if they are to feed and swallow while in a confined space. Long legged birds, such as storks, cranes, and herons, must be able to stand upright; they must never be confined with their legs folded.

Too much room, though, can be a detriment. When confining a panicky bird, it must not be given so much room that it injures itself by bashing against the walls of its confines. Instead, it should be given a minimum amount of space, just enough to comfortably turn around. Birds that struggle continually will tend to overheat and must have adequate ventilation to prevent hyperthermia.

Choosing the appropriate size and shape of temporary housing, as well as what material it is made out of, will depend on the species and the situation. As a rule, wild birds should not be housed in wire cages as they can damage their feathers. See-through plastic or glass containers should be avoided.

Plastic "airline kennels" come in many sizes and provide excellent airflow. While these modular cages are suitable as temporary housing for a variety of birds, they require modifications to house certain species. Loons, grebes, herons, and pelicans have been known to damage their bills on the metal grate openings. To reduce the risk of injury, the grates should be lined on the inside with shade cloth. At the very least, a pillowcase or sheet should be draped on the inside to prevent birds from damaging their bills on the bars.

The floors of these plastic carriers are slick. Towels can be used to provide traction and cushioning, and to absorb excrement. Anti-fatigue, anti-slip flooring comes in various dimensions and can be cut to fit the bottoms of the kennels. This material can be washed and disinfected for reuse.

Certain birds may be more comfortable standing on a platform so they do not bend or break their tail feathers. Cormorants are a good example.

A secure platform can be made out of a 4 × 4 inch (10 × 10 cm) piece of wood, 12 inches (30 cm) long or so, with two footers to keep it from flipping. It can be wrapped in a towel or anti-slip material and set in the center of the kennel.

For smaller perching birds, a washable and reusable perch can be made out of PVC. PVC footers filled with sand will keep the perch from flipping. The perch can be wrapped with sisal rope or vet wrap, padded as needed. Perches may not be suitable for newly captured birds that are panicked.

Collapsible pet carriers provide good ventilation and can be used for a variety of species. The plastic or waxed boxes are preferred over the cardboard ones, as they remain sturdy under wet conditions and can be cleaned for reuse. Towels or rags should be placed on the floor of the carriers to keep animals from slipping. These carriers are not suitable for cormorants, as they don't provide enough headroom.

To utilize plain cardboard packing boxes, they must first be modified to allow for good ventilation. A minimum of eight air holes on each side of the box should allow for sufficient airflow. The holes should be about one-quarter the size of the bird's skull – approximately 1 inch (2.5 cm) in diameter for large birds. A word of caution: cardboard boxes will start to fall apart if they become damp; care must be taken that they do not become wet.

Deep, rectangular boxes and bins can make excellent housing for loons and grebes, allowing handlers safe access to the birds from above. Plastic storage bins can be modified to house a variety of species as long as there is adequate airflow. Depending on the species the bins are intended for, appropriately sized air holes can be drilled along the sides.

Diving birds, such as loons, grebes, auks, and sea ducks, require special consideration during confinement. These birds are built for life on the water. They are like boats, with a keel at the front and "rudders" at the rear (Fig. 53). Subsequently, they have trouble standing and walking on hard surfaces. When beached, these species are usually found recumbent, with their keel supporting most of their weight. With almost nothing protecting the keel bone, these birds quickly develop what are called "keel sores." These species will also begin to develop leg and foot problems from resting on a flat surface, which forces their legs into a hyperflexed state.

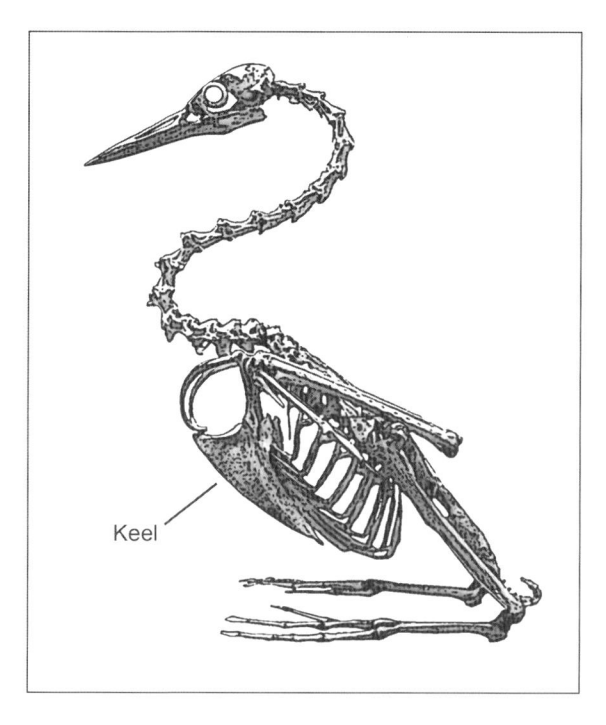

Fig. 53 Diving birds, such as grebes, are built for life on the water.

In 1988, aquatic bird specialists International Bird Rescue (California) developed a net-bottom caging system that helps delay the onset of these serious and often fatal ailments (Fig. 54). The flexible fabric distributes a bird's weight off its keel. The netting also helps prevent feather rot by allowing excrement to fall through, away from the bird. Smaller versions of these can be made out of PVC and sized to fit snugly inside plastic carriers, bins, or boxes.

In the absence of a net-bottom cage or insert, plush, uneven padding may be provided as a very temporary solution. A thick layer of balled-up newspaper covered over with a large towel will help distribute the bird's weight but will not do much to address injury to the leg.

In 1995, International Bird Rescue developed a tool to further address development of keel sores and leg ailments in aquatic species. Fashioned after a pressure relief cushion, the "doughnut" takes pressure off the keel entirely. The U-shaped cushion elevates the bird slightly, delaying the onset of hock joint ailments (Fig. 55). Use of this tool is normally restricted to the rehabilitation setting as it is labor intensive and requires skill to apply without harming the animal.

In captive environments, panicked wild birds will have a tendency to run straight into the sides of their enclosures, causing injury to their "wrists." In the

Fig. 54 Example of a net-bottom caging system.

Fig. 55 A soft "doughnut" is used to take pressure off a bird's keel.

mid-1990s, International Bird Rescue came up with the concept of soft-sided caging systems that greatly reduce the severity of these "bumper wounds," if not preventing them altogether.

Soft-sided caging systems can be built to suit the species being housed. Constructed of tarpaulin, or canvas-like material, these modular tent-like systems are relatively simple to build and can be cleaned and stored for future use. Petite versions can be constructed out of PVC rather than wood framing. Soft sides can also be used with net-bottoms for housing active aquatic birds such as scoters or other sea ducks.

Taking this a step further, the use of round cages, rather than square or rectangular, can further reduce the chance of birds striking the walls of their enclosures straight on. This design also minimizes the chance that birds will become "stuck" or crowded into corners.

Some species, such as American coots, will have a tendency to jump up repeatedly. Ceilings of hard-sided carriers can be fixed with soft foam padding to prevent trauma to a bird's head. Foam padding, however, is typically not advisable for caging, as it can be a breeding ground for bacteria. If used to soften the sides or ceiling of a cage, it must be kept clean and dry.

Certain materials should be avoided altogether as substrates and bedding for temporary confinement. Wood shavings, hay, and straw should never be used to house birds. Aspergillosis is a serious avian respiratory infection caused by a fungus associated with dusty straw and moldy hay. Dust and small particles from wood shavings can also cause respiratory issues and eye problems.

Towels and sheets can be used to line the bottom of cages. If more padding is needed, shredded or balled-up newspaper or foam padding can be placed underneath the towel or sheet, as long as it stays dry. Anti-fatigue mats can also be used for larger birds, such as pelicans.

9 Capture, handling, and confinement of land mammals

Techniques for capturing wild mammals

Whether one is out to net a pocket gopher or a coyote, the mechanics of wielding and scooping with a net will be similar. When approaching an animal, the capturer will want to keep a low profile, holding the net low, approaching close enough that the hoop can be extended beyond the animal's head, anticipating its direction of travel and speed.

The actual act of netting the animal should be a single swift move. Ideally, a capturer will stay coiled – elbows at their side (or slightly back) until within "striking" range. They will want to aim so that the head of the animal is in the center of the hoop.

Once the animal has been netted, the hoop should be lowered and held flush to the ground. To prevent the animal from escaping, the hoop can be spun a couple of times to twist the opening shut. The tip of the hoop can be used to hold it closed.

For tighter restraint of a netted animal, the hoop can be drawn over the animal's body and down, with just enough pressure as needed (Fig. 56). This can be a useful method of retraining an animal for an injection. A word of caution though: animals must not be kept under this type of physical restraint for long, as they might overheat.

When netting an aggressive animal, herding boards can be used to keep the animal secured in the bag and away from the netter. If using an open-ended net, the herding boards will also be used to encourage the animal into the carrier.

In the absence of a hoop net, another simple method of containing a small animal is to cover it with a container such as a jam jar, coffee tin, cardboard box, or plastic bin. Once secured, a thin piece of material can be carefully slid underneath. The animal can then be transferred to a more appropriate container.

Another basic strategy for securing a loose animal is to use barriers to reduce the space it has. Simple materials such as cardboard or tarpaulin can be used to construct chutes and funnels to drive or block its movement, making it easier

---

Wildlife Search and Rescue: A Guide for First Responders, First Edition. Rebecca Dmytryk.
© 2012 John Wiley & Sons, Ltd. Published 2012 by John Wiley & Sons, Ltd.

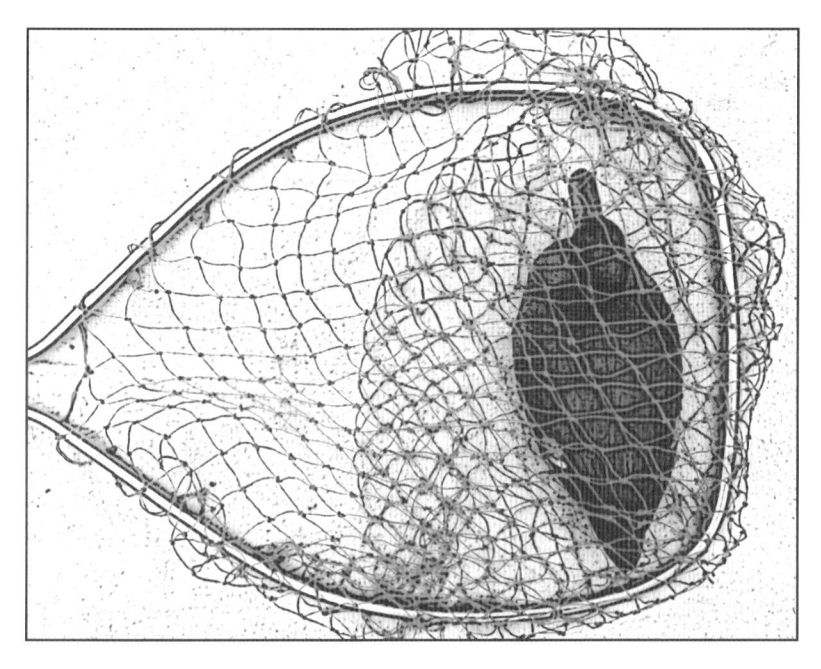

Fig. 56 With the animal in the bag of the net, the leading edge of the hoop can be drawn over the animal, restraining it further.

to contain. Walk-in-type traps should be considered as a means of drawing an animal into an enclosure where it can then be netted.

Chemical immobilization

The use of chemicals to slow or immobilize an already compromised animal is extremely risky and should be considered as a final option, after other possible methods have failed.

Special circumstances and particular methods

Small rodents

Mice, rats, squirrels, and the like can be extremely swift. Rescuers may have the tendency to want to try capturing a fast-moving rodent with a hoop net. The problem with this method is that there is a risk of the animal being struck and injured by the hard rim of the net.

The hoop shape can also be a problem when trying to contain a speedy animal against a squared-off surface, such as a wall. The use of a square-headed net, similar to ones used for snakes, will eliminate gaps that the animal could squeeze through.

Unless alarmed, mice and rats (and other small animals) will tend to gravitate to and track against a vertical surface, such as a wall. This thigmotactic behavior can be predictable and can help rescuers build a successful capture strategy.

One method would be to place a square-headed net against a "wall," making it appear like a safe place to hide. With light pressure, the animal can be encouraged to run to and along the wall and into the net. Where there are no vertical surfaces, they can be constructed of wood or cardboard. This funneling concept can also be used with larger mammals.

Large rodents, porcupines, beaver

Netting a porcupine, even with a fabric bag, will require additional handling of the animal and is therefore not as safe a method as herding. Even in small mesh bags, their quills can become snagged as the animal struggles.

Using tarps, sheets, plywood, or cardboard, porcupines can be herded into a confined space and into an appropriate carrier, such as a modified pipe or drum. Brooms and herding boards can be used to gently encourage the animal to move forward. Porcupines should never be placed into wire cages. As for beaver, herding should be attempted first, with netting of the animal as the second option.

Lagomorphs, rabbits and hares

Like deer, rabbits and hares are prey species and will have an alarm reaction so severe they may break their own spine. In being particularly careful when capturing and handling these animals, rescuers must minimize auditory and visual stressors. They may be best guided through a chute into a carrier. Spotlighting at night may be useful in capturing them.

Xenarthrans, anteaters, armadillos

Once considered members of the order Edentata, anteaters, sloths, and armadillos are now classified in the super-order Xenarthra. These are some of the oldest and oddest mammal species on earth. As such, there are a few peculiarities that rescuers must consider for safe capture and restraint. Their mouths are delicate and care must be taken not to injure them. A padded hoop net with soft rims can be used. A better approach may be to use herding techniques. Anteaters and armadillos have powerful legs and claws used for digging. Rescuers should wear leather gloves for protection. Xenarthrans are also heterothermic, meaning their body temperature is influenced by their environment.

Skunks

Skunks are not typically skittish. They will tolerate being herded, even netted, quietly and slowly. One method of confining a skunk is to offer it a place to hide – preferably an enclosure designed specifically for skunks.

A large drainpipe can be modified to make a skunk carrier. Each end can be fitted with a wooden frame and guillotine door. Once inside with the doors closed the skunk may still spray, so precautions must be taken to avoid contamination.

If a skunk must be netted, rescuers can shield their advance using heavy sheets to take the brunt of the skunk's spray. Once netted, the sheets can be placed over the skunk to reduce stressors. Long-handled, deep nets have been used to carry skunks without them spraying.

Canids

The Collarum™ is a snare-type trap designed exclusively for the live capture of canids (coyote, fox, dog, wolf, etc.). The trap is triggered by a tug or pull action rather than the standard push or depress action, making this trap canine-specific.

When triggered, a cable loop flips up and is then cast over the animal's head and neck. This dual action works well when the animal approaches from straight on. However, there is a risk of injury if the animal approaches from the rear or either side. Therefore, careful placement of the trap is critical. The animal must only be able to reach the bait and spring the trap directly.

Once the dog has been snared, a locking device on the cable prevents the animal from being strangled. While this device is designed to capture wild canids alive, they can severely injure themselves while struggling to get free. Rescuers must never leave this trap unattended. They can keep watch, monitoring the trap from a distance or remotely so that the animal can be netted and removed from the snare within minutes.

Deer

An alert and mobile adult deer may be one of the most dangerous animals a rescuer will face. They are highly nervous animals that can react violently when alarmed.

When frightened, they can become predictably unpredictable and dangerous to themselves and others, crashing through fences or windows, and leaping to their death in the attempt to flee. When netted, they will struggle and risk serious injury.

If presented with an adult deer that is still mobile, great caution must be taken when planning and implementing a recovery. The safest option, for both deer and human, will probably be sedation of the animal prior to any handling or restraint. Anesthesia must be administered or supervised by a veterinarian experienced with sedating wildlife.

One method for securing small or manageable deer is with a large stretch of netting material. In an ideal situation, the deer will be cornered. Two or more rescuers, each holding an end of the material, will walk towards the animal. As it attempts to flee between them, it will walk into the netting. Once in the material, the rescuers will quickly work to keep the deer from struggling in the net.

Working from the backside of the animal, small-sized deer can be restrained at the shoulders and rump. It is extremely important to cover the deer's eyes using a towel. Downward pressure on the neck will help keep its head down. Antlers that are not in velvet may be held onto, but should be covered with towels. Antlers in velvet must be protected from being damaged. Deer must not be restrained for long as they are highly susceptible to capture myopathy. If the animal is to be transported, it should be loaded into a suitable deer carrier without delay.

Another tactic for unapproachable deer is to use a downward facing net launcher, using bait to draw the animal into the capture zone. This should only be considered as a last resort as deer can become seriously injured when they struggle in netting material.

To minimize handling, rescuers should consider options, such as using a series of chutes to slowly direct or "squeeze" a deer's movement.

Fawns are much less risky to handle. They can often be captured and carried safely in a large blanket. Care must be taken that they do not overheat while wrapped in any sort of material. In general, a fawn can be transported safely in a well padded dog crate.

Physical restraint of land mammals

Bats

Rescue personnel should always wear gloves when handling bats unless directly instructed otherwise by a bat specialist who is supervising their activity.

Nitrile gloves are often used to restrain and examine very small bats. They provide the carer with a protective barrier from potentially infectious body fluids while providing tactile sensitivity. Thicker gloves are necessary for larger species.

A bat's wing consists of bones that are similar to those in a human arm and hand. Essentially, a bat's wingtip is the extension of its third, or middle, finger. Their thumbs are not incorporated into the wing itself, but extend above its leading edge and are used to cling and climb. Bats can sustain serious injuries if they are pulled forcefully from the surface upon which they are clinging. Instead, rescuers must take the time it takes to gently relieve the animal of its grip. With a gloved hand cupping the animal, a playing card can be used to softly encourage the animal to release its hold. As it does, the animal can be lifted up and away so it does not reattach to the roosting surface.

Some species, such as the crevice-dwelling bats, can be gently restrained in a closed hand exerting minimal pressure. Holding a bat too tightly can cause it to panic. Holding a crevice-dwelling species too loosely, however, may prompt an escape attempt. Foliage-dwelling species can be held relatively loosely on a cloth or gloved hand or restrained ever so gently between the thumb and one finger (Fig. 57). Bats must never be pulled or held by their wings.

Small rodents

Small rodents can be handled with modest leather gloves, supple enough to safely manipulate the animal and strong enough to protect against bites. They can be secured within a clenched hand. For more extensive handling, smaller rodents may be gently but firmly scruffed. As needed, the base of the tail may be held for better control and examination.

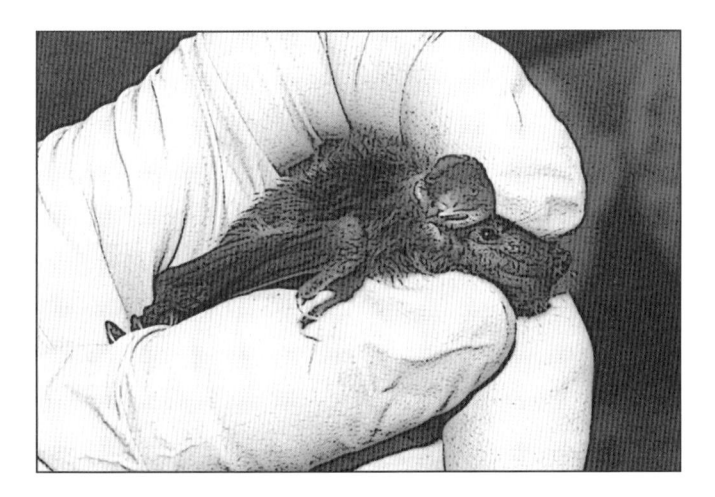

Fig. 57 A method of restraint for foliage-dwelling bats.

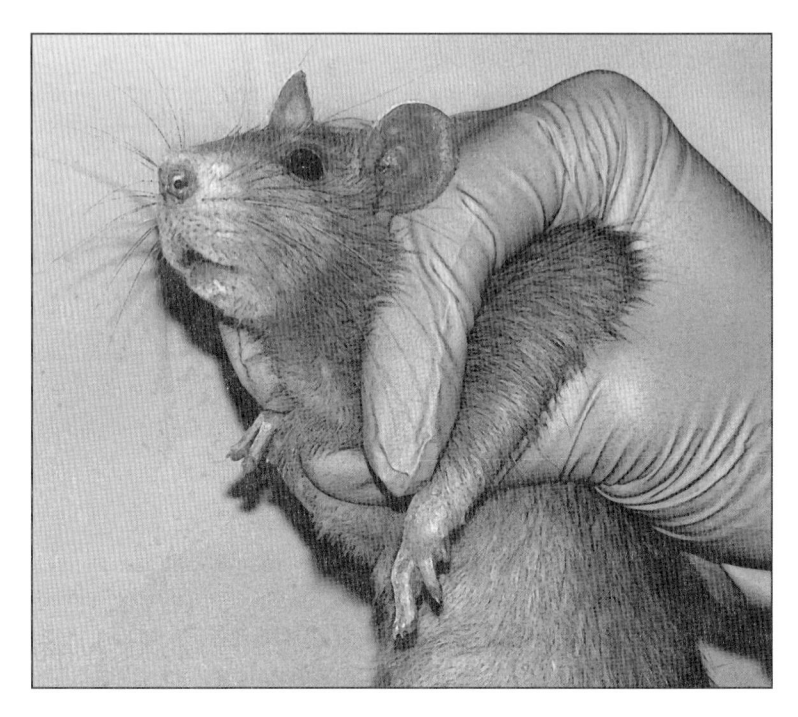

Fig. 58 A method of restraint for some rodents. Courtesy of Bat World Sanctuary.

Some rodents can be held with their head resting between the index finger and middle finger, with their forelimbs secured by the thumb and ring finger (Fig. 58). An alternative to this is to hold the head between the thumb and index finger. Obviously, care must be taken with the amount of pressure used with these methods – again, the least amount of force to get the job done without injuring the animal.

Another way to secure a very small mammal is in a soft plastic or fabric handling cone or tube. This type of restraint can make an examination of a wriggly rodent much easier for the animal and its handlers.

Talpids, moles and relatives

Moles have surprisingly strong forelimbs that they will use to try to squirm from a hand grasp; therefore, they are best restrained by the scruff, with the other hand supporting the base of the body.

Shrews are extremely susceptible to stress from handling. Although they may be restrained by the scruff, it may be best to encourage them into a plastic or fabric tube with plenty of ventilation.

Squirrels

Squirrels have long incisors and powerful jaws. They are also high-strung and susceptible to shock from handling. Handlers can minimize the risk of being bitten and dramatically reduce stress on an animal by restraining it in a fabric cone.

A fabric handling cone can be made from lightweight durable cotton, such as a denim or canvas material. Strips of Velcro can be used to keep the cone sealed while allowing the handler easy access to various portions of the animal's body.

Another tool for safer restraint of squirrels is a cone-shaped collar made out of a plastic funnel. The end of the funnel can be cut to the appropriate size and then split longitudinally.

To apply the collar it is recommended that the animal be first placed into a canvas handling bag with only its head exposed. The collar can then be gently slipped around the squirrel's neck and taped closed to a snug fit.

To manually hold a squirrel, the handler, with gloved hands, will want to gain control of the head at the neck using his/her thumb and index finger. The other hand will be used to gain control of the rear legs and tail while also providing support to the body. Squirrels must never be grabbed or picked up by their tails.

Opossums

Opossums, even very young ones, will exhibit threat displays such as opening their mouths, showing all their teeth. They may also hiss, lunge, and drool. Another defense mechanism is their ability to feign death – an involuntary reaction to extreme fear. In this state, an opossum's eyes may appear glazed and its mouth might be fixed in a grimace.

To restrain an adult opossum, the handler, wearing heavy leather gloves, can gain control of the animal's upper body at the nape of the neck, with the other hand providing support of its body near the base of the tail (Fig. 59). A juvenile opossum can be restrained by placing its head between forked fingers with the base of its tail and rear legs restrained as needed. Opossums must never be picked up or carried by their tails.

Porcupines

Porcupettes and smaller species of porcupine, such as the prehensile-tailed, can be safely handled using thick leather gloves. North American species do not have quills on the undersides of their tail and abdomen and can be safely restrained by an experienced handler. Quills must never be grabbed to hold on to or used to move a porcupine as their skin can easily tear.

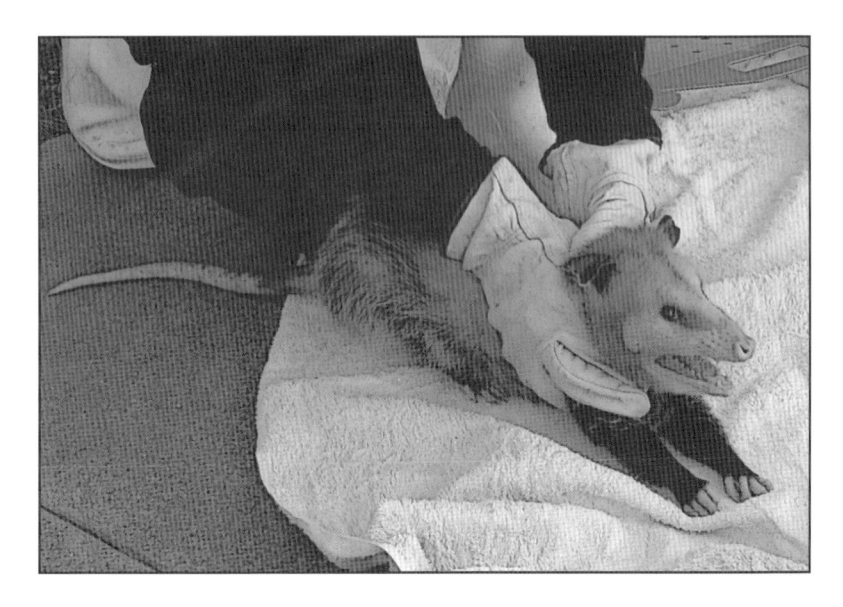

Fig. 59   Restraining an adult opossum.

Lagomorphs, rabbits and hares

Rabbits and hares, with their extreme alarm response, can become injured during handling and restraint. It may be less stressful and safer for a rabbit to be quickly confined in a pillowcase or fabric sack, making sure it has sufficient air and does not overheat. The animal's body should be supported while it is in the bag, held against the handler's body when it is safe to do so.

To physically restrain a rabbit, the handler will start by gaining a grip at the nape of the neck and applying just enough downward pressure so that the animal cannot kick back and up; this is how they can injure themselves. Using the other hand, the handler will either support the rump and bring the animal's body to rest against their own, or "sandwich" the rabbit by supporting at the abdomen with the animal's feet facing out. Rabbits or hares must never be picked up by their ears.

Small and medium-sized carnvores

Hands-on restraint of conscious and alert adult badgers, weasels, otters, raccoons, and bobcats is not recommended without additional equipment such as nets, netting, squeeze cages, or handling bags.

Most raccoons can be baited into a trap or cage. Because they are nocturnal, such capture attempts will probably have to take place at night.

If a catchpole will help rescuers extract or move a raccoon, the loop must be slipped over the head and one shoulder – an animal should never be pulled by the neck and must never be lifted by the neck without also having its body supported. This method can also be used to control badgers, otters, and other mustelids.

Mustelids, badger, otter, weasels

Even the smallest members of the weasel family will be scrappy. Precautions must be taken when handling these carnivores to prevent bites to humans and to prevent the animals from damaging their teeth. If a hoop net is used, the hoop itself should be padded, and wire caging must be avoided.

Larger species can be initially controlled using a catchpole. The noose can be placed around the neck and one shoulder if possible. Animals must never be lifted in this fashion but can be carried and placed into a suitable container with the help of a second handler at the rump. Further restraint should be done under anesthesia or using a squeeze bag or cage.

Skunks

If a skunk must be manually restrained, it can first be netted and then covered with a sheet or large towel, thereby preventing it from seeing the handlers. If the skunk discharges its noxious spray, the coverings will, it is hoped, take the brunt of it. Once the skunk has been shielded, downward pressure can be applied using another net or shield, immobilizing the skunk long enough for a handler to get into place. Using heavy-duty gauntlet-style gloves and a face shield, the handler can then attempt to gain control of the head, gripping the base of the skull, while at the same time tucking the tail under the skunk's body, reducing the chance that it will spray.

Coyotes and foxes

Coyotes and foxes can be scruffed – held tightly by the nape of the neck using gauntlet-style gloves. Although they will probably try to bite at first, many times they will submit while being held by the nape, going limp, just waiting for the second they feel a loosening of the handler's grip. If the animal must be lifted while being scruffed, the handler will want to support the back end at the rump.

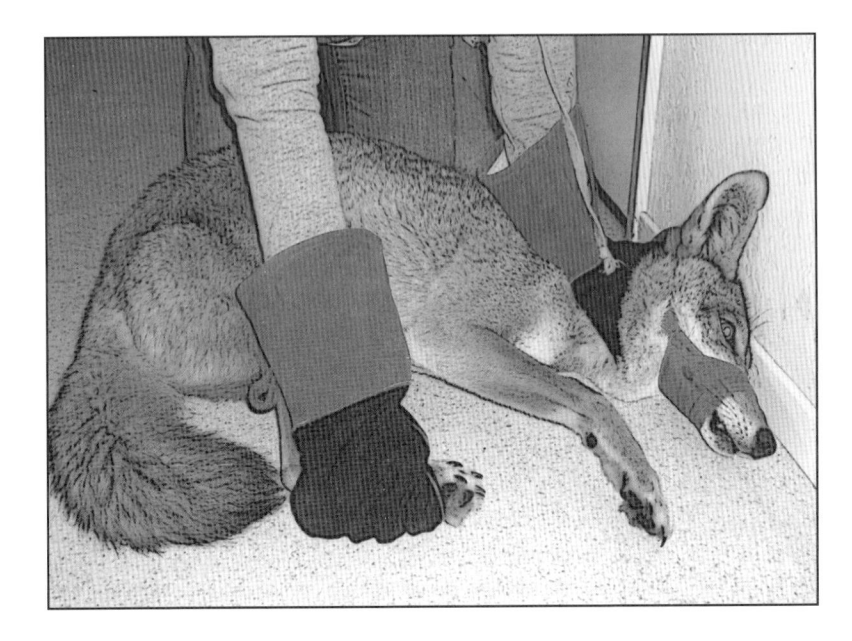

Fig. 60 Restraining a coyote or fox.

A coyote or fox must never be pulled or lifted by its tail. They can also be restrained at the nape with their legs held securely (Fig. 60). Muzzles can be used on wild canids, but for very brief periods only, as the dog must be allowed to pant for it to stay cool. Canids may also be controlled using a catchpole but must never be lifted by this method alone.

Felids

Handling and restraint of alert and mobile adult cats can be accomplished most easily with use of an open-ended hoop net and herding boards. From there the animal can be transferred into an appropriate carrier. Snare poles are not advised for use on cats. Small kittens can be scruffed using gauntlet-style gloves for protection.

Processing mammals from nets and cages

Once an animal has been trapped or netted, rescuers are faced with the task of transferring, or processing, the animal out of the net and into a suitable carrier. This is reasonably simple if the animal has been captured in an open-ended net where there is little or no hands-on contact. However, the following methods are presented to give handlers suggestions on how to remove mammals from standard hoop nets and traps.

One of the most basic methods of removing an animal from a hoop net begins by throwing a towel over its head and body. This reduces visual stress, applies a bit of downward pressure, and offers rescuers an added layer of protection.

With the towel over the animal's head and shoulders, the handler, using gloved hands, will want to gain control of the animal's head, typically with their thumbs on the back of the skull and his/her fingers against the angle of the lower mandibles. If they are not sure where the head is, they may want to apply slight downward pressure – just enough to restrain the animal's movement, until it becomes oriented.

To remove the animal from the net, a second handler will help peel back the material, exposing the animal's body and replacing the first handler's grip with his/her own. This coordinated effort should be performed methodically and the handlers should communicate clearly with each other as they exchange grip of the animal.

Once the animal is securely restrained outside of the net it can be manipulated into a carrier. Customarily, a carrier is placed directly in front of the restrained animal. Sometimes the lip of the carrier can be placed under the animal's chin. The animal is guided into the carrier, taking care never to lift the animal by its neck. Most of the momentum is through pushing and lifting of the animal's body. When the chief handler is ready to release his/her grip on the head, another handler will want to hold the carrier closed while his/her quickly slips his/her hands out.

Another way to get an animal from a hoop net into a side-loading carrier is to place the hoop of the net tightly against the opening of the carrier (Fig. 61). Gloved hands or herding boards can be used to encourage the animal into the carrier.

Getting an animal from one container to another can be extremely difficult. If possible, the two containers should be joined together. Large towels, sheets or a tarp can help seal any gaps between the carriers. The cage that the animal is to move into should be covered so that it is dark inside. This will encourage the animal to move into it, seeking cover.

After joining the two cages, animals should be shielded from seeing or hearing humans and given a few minutes of quiet before being urged to move into the new container. It may take some finesse to get an animal to budge, but sometimes the lightest touch with the bristles of a broom can get them to move out. Unfortunately, though, transfers often require that animals be wrangled.

To simplify transfers and eliminate additional handling, the author has devised a tool specifically for cage transfers. It is a chute made from durable nylon material. It attaches snugly and securely to both containers, forming a tube. The chute is modifiable, easily transforming into a handling bag for safer extraction of animals from cages and safer, less stressful, exams. The chute comes in various sizes.

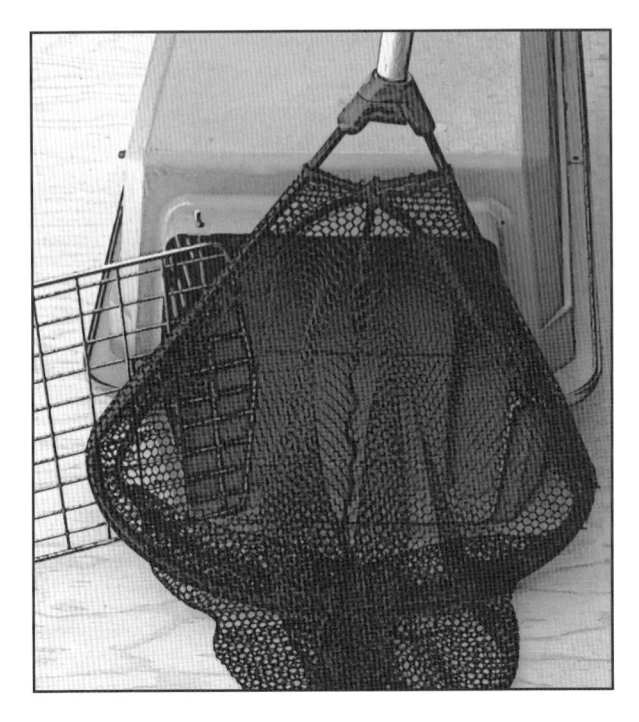

Fig. 61   Transferring an animal from a hoop net into a side-loading carrier.

Temporary confinement of land mammals

Before an animal is captured rescuers should prepare for it to be safely and securely housed. No animal should have to wait while a cage is set up – this must be done ahead of time to reduce stress. The size and material will vary depending on the species and its physical condition.

Once captured, wild animals that are strong enough to fight for their freedom will do anything they can to escape. If an animal is to be housed for a few hours, or overnight, rescuers must prepare to offer it a clean, dark, ventilated, temperature-controlled environment in which it cannot cause itself greater harm. For example, wild animals can injure themselves on wire; therefore, solid wire caging is not recommended when other options are available.

Animal carriers come in many shapes, sizes, and materials. Temporary housing can also be made from common household items. In selecting or constructing suitable confines, a few basic principles must be considered.

As for the space, temporary housing should provide the animal enough room for to maintain normal body posture and to turn around without hitting the walls.

If wild animals are given too much room, especially high-strung species, they can seriously injure themselves. Too much space can also make retrieving them more difficult. The animal's condition and its ability to thermoregulate must also be considered when deciding on the size of caging. Newborns, for example, need just enough room to curl up in, similar to what their nest would be. If they are given too much room to squirm about, they can quickly become chilled.

Ventilation is another important consideration when deciding on temporary confinement. Again, the animal's condition and level of activity must be taken into account. High-strung animals will require greater air exchange to expel the heat they generate, while newborns will need less air exchange to retain warmth. As a rule, there should be air holes, approximately 1/4 inch (6.25 mm) in diameter, on at least two sides of a carrier. Larger, more active animals may require greater airflow through bigger and more numerous openings of 1 inch (25 mm) diameter or so.

Regarding flooring, most solid materials will feel slippery to an animal. Bottoms of cages should be lined with something to give traction, such as a towel or a section of an anti-fatigue mat.

In addition to the floor covering, some sort of bedding or hide should be provided, such as a good-sized towel that the animal can curl up under. Regarding the material, terry cloth should be avoided when housing small animals because their tiny toenails will snag on its threads. Tighter weaves, such as flannel, are safer. Synthetic materials, such as fleece, can build up electric charge. The sparks and shock from this static electricity can be alarming and painful to a wild animal. Natural fabrics made of cotton, rayon, or wool are less likely to spark.

Regarding the material making up the confines, rescuers must consider how an animal will try to escape. Will it try to gnaw, chew, or dig its way out? Will rescuers be able to control the temperature inside? The material used must be able to hold the animal's weight and withstand soiling without falling apart. Wild animals housed in glass and clear plastic containers may exhaust themselves trying to get through. This can be fixed by making the walls appear solid by applying paper or fabric to the exterior.

Very small newborn mammals, up to a few inches in size, can be safely confined in quart-sized (950 ml) plastic or tin containers. Air holes should be drilled or punched through the lid. Fancier models can have screen fixed to a larger opening in the lid. Pet stores offer something similar – small, clear plastic animal tanks with pop-up screen lids.

Bats can be temporarily housed and transported in soft-sided mesh carriers. It is important to line the bottom of the carrier with padding, such as foam or layers of tightly woven fabric, so that if a bat falls to the floor it is less likely to be injured. Crevice-dwelling species should be provided places to hide inside the container, such as folds of draped non-snagging fabric. Crevice-dwelling species can be housed together, while foliage-dwelling species should be confined separately.

Another method for confining mammals that are approximately 8 inches (20 cm) long or so is a clean, well ventilated 5-gallon (19 litre) screw top plastic pail. Air holes must be drilled in the sides and the top for adequate ventilation. The bucket can also be tipped onto its side and stabilized with footers to keep it from rolling. This simple design works nicely for short-term confinement of small mammals such as young skunks.

Similar specialized caging for larger skunks, and badgers, can be fabricated out of PVC pipe. The section of pipe should be at least 12 inches (30 cm) in diameter by 24 inches (60 cm). Both sides of the PVC tube should have at least eight air holes, approximately 1/2 inch (1.25 cm) in diameter. Plywood framing keeps the pipe stable. Guillotine-style doors can be added on both ends, making this a versatile caging system.

Deep plastic or rubber bins, even trash cans, can be modified to temporarily house a variety of species from ground squirrels to opossums. The air holes should be situated high, out of the animal's reach, especially if the animals will try to gnaw or chew their way out.

One of the advantages of using tubs or bins, even deep cardboard boxes, is the top access. It is often easier and safer for handlers to reach animals from above than reaching in through a side door.

Collapsible pet carriers made out of cardboard, waxed cardboard, or corrugated plastic can be suitable for a variety of small mammals as long as the animals cannot escape through the air holes.

Plastic "airline kennels" come in many sizes and provide excellent airflow. These modular cages are suitable for a variety of mammals, including raccoons, bobcats, and coyotes.

Deer are extremely sensitive to the stress of capture, handling, and confinement. To reduce stressors they should be housed in a dark and quiet environment. A fawn can be housed in a large plastic airline kennel with straw as bedding. Juveniles and adults can be temporarily confined and transported in a specially made deer box. Deer boxes can be made out of plywood. They should be designed to give the deer enough room to stand up or lie down but so narrow that they will discourage a deer from trying to turn around. The boxes should have air holes cut out towards the top and at the base of each side. The holes can be 1 inch (2.5 cm) in diameter on both sides. Each end of the box should be fitted with a guillotine door. A thick layer of straw can be used for bedding.

10 Capture and handling of reptiles and amphibians

Most amphibians can be captured and manipulated using soft, fine mesh nets, like those appropriate for tropical fish. When handling amphibians, it is important that hands be absolutely clean, preferably wet. If using gloves, they should be either Nitrile or powder-free latex and should be rinsed before coming into contact with the animal.

Most amphibians can be cupped or held by placing the animal's neck, gently caught, between two fingers with the body completely supported, similar to grasping a songbird.

If a reptile is hiding, lures can be used to entice it into the open. In addition to food bait, warmth from a heating pad or heat lamp can help draw a reptile out and even into a container.

Lizards are often easiest to capture using a net, although there is some risk of injury should they receive a blow from the net's rim. A small catchpole, using monofilament line for the noose, has been used to successfully capture lizards. Snakes with spade-shaped heads can also be captured using a noose; however, this technique is not recommended if other options are available. Venomous snakes do not have as strong of musculature as constricting species do, so they must never be lifted by the neck alone, without supporting their body as well.

Snake tongs are excellent tools when snakes are cold or cooperative. Warm, active snakes, though, can be extremely fast and difficult to grasp. A square-headed net can work very well for capturing venomous and non-venomous snakes. The sock should be deep and made of soft, knotless, small mesh. The bottom can be modified even further to allow it to be cinched and separated from the rest of the net, forming a sack for carrying non-venomous snakes.

Venomous snakes should not be confined or transported in netting or fabric. Once netted, the snake should be placed into an appropriate container that can be locked. Deep plastic bins with screw down lids can be modified with air holes to contain venomous snakes.

11 Marine mammal rescue

Rescuing seals and sea lions

A pinniped is a member of a group of "fin-footed" marine mammals classified under the order Pinnipedia. This includes the families Otariidae (sea lions and fur seals), Phocidae (true seals), and Odobenidae (walruses). Each family is distinguishable by its unique physical appearance and locomotion.

Pinnipeds are basically land mammals that forage in the sea. It is normal for them to come ashore to get warm and dry and to rest. Common injuries include fishhooks, fishing net entanglements, propeller strikes, shark bites, stingray barb wounds, and domoic acid poisoning. Because each species differs in appearance, and when approached and handled during a rescue, it is therefore imperative that rescuer personnel be familiar with the species they may encounter.

The otariids, or "eared seals," can be quickly identified by the presence of external earflaps. They are also much more agile on land than other pinnipeds. Being quadrupedal, sea lions and fur seals are able to clamber atop rocks and scale cliffs. Other distinguishing characteristics include their fore flippers, which are large and wing-like. Close inspection of their rear flippers will reveal long, fleshy, cartilaginous digits extending beyond each toe.

Otariids are gregarious and may congregate together in large groups, often resting on top of or up against one another. This demonstration of positive thygmotaxis is sometimes informally referred to as "gettin' thiggy."

Another behavior unique to otariids is called "jug handling." This is when they rest, floating at the water's surface, and raise a hind flipper and fore flipper above the water. When the appendages meet in an arc, it can resemble the handle of a jug. Otariids can rest in this position for a relatively long time, occasionally eliciting concern from passersby.

Otariids are like the dogs of the sea. They are extremely intelligent. Of all the otariids, perhaps the California sea lion is the most familiar. With their ability to adapt to captivity and be trained, California sea lions are used widely in entertainment, research, and military operations.

Along urbanized coastlines, where sea lions are accustomed to humans, some may allow rescuers to approach within a few yards (about 3 m) without displaying concern. That said, if a sea lion suspects a bad encounter it will be very quick to flee for the safety of the water.

In most cases, when it is clearly called for and safe to do so, rescuers will want to get between the animal and its escape route – in this case, the water. If the sea lion is alert and appears agitated, rescuers may want to use the herding boards to conceal their approach. Once in position between the animal and the water, the netter and herders will move in quickly. Some sea lions will make a dash to the water, some will hold their ground, others may charge.

Members of the family Phocidae are considered the "true seals." They are also referred to as the "earless" or "crawling seals," distinguishing them from their more mobile cousins. Besides the absence of external earflaps, phocids have flippers that are covered with hair, with claws at the ends of the digits. Their locomotion can be described as more like that of an inch-worm.

Some phocids, such as the harbor seal, can be very timid and skittish, while others, such as the elephant seal, tolerate encroachment.

Elephant seals can be described as being slate gray to a soft golden color with a lighter underside. They go through what is called a catastrophic molt, where they shed their fur. This can drive them ashore.

Elephant seals are even more unusual in that they are the only species of seal that will actively toss sand on their backs. White discharge from their nostrils is normal, as is occasional snorting. Young are weaned abruptly at about four weeks.

Young seals and sea lions

Harbor seal pups nurse for about a month. They can swim at an early stage but are often left on land while the mother forages. The presence of an umbilical cord indicates the pup is newly born. A healthy harbor seal pup will appear bright and alert. It should be well fleshed and have a mask of wetness around its eyes, indicating that it is well hydrated. If it is healthy, it may simply need to be left alone for its mother's return. The adult harbor seal is very shy and will not return to her baby if there are people or other predators on the beach. Unless the rescuers on scene are permitted to rescue marine mammals, they will want to alert the nearest marine mammal rehabilitation center of the pup's exact location.

California sea lions give birth in colonies on offshore islands. A newborn pup will be about the size of a big cat and will bleat like a baby goat. Typically, after bonding with her pup for a few days, the mother will leave it from time to time while she forages in the ocean. The pup is left in the company of other pups and adults. Pups begin to wean at six months but might stay with their mothers for a year or longer.

A California sea lion pup found on the mainland with no mother in sight is likely to be orphaned – the mother may be ill, suffering possibly from domoic acid poisoning, or dead. To date, there has been no successful rehabilitation of newborn California sea lion pups. No one has yet found a way to raise them without them imprinting on their human caregivers.

If the mother is present, and appears healthy, there is a chance, though slim, that the pair can be relocated to their island home. If she has just given birth and it is safe for her to remain where she is, she must be left undisturbed so that she can bond with her baby and it can nurse. The task of securing the mother and pup will be left to those trained and permitted to take action on behalf of marine mammals.

When responding to a beached pinniped, the rescuer will want to first assess a beached pinniped's condition, observe its body posture and behavior from a concealed location if possible. Signs of distress include dry, gummy eyes, prominent hipbones, spine, or ribs, coughing or choking, disorientation, and unresponsiveness. If closer evaluation is necessary, a single rescuer may approach slowly and quietly, halting before their presence elicits a response.

Often, a beached pinniped will draw a crowd of bystanders. Rescue personnel should be prepared to handle the public, keeping them at a safe distance of at least 50 feet (15.24 m) away from the animal so that it will not feel pressured.

In general, a rescue of a pinniped can be broken down into three phases – capture, restraint, and confinement. In each phase the rescuers will select the appropriate equipment and technique based on the animal's size, weight, liveliness, and temperament. Because pinnipeds will have a tendency to overheat during capture and transport, the appropriate equipment, caging, and transport should be assembled and ready before the animal is caught.

As for marine mammal capture equipment, rescuers will want to have thick leather gloves, a couple of heavy-duty, gauntlet-style welder's mitts, an assortment of bed linens, bath towels, herding boards, a few heavy-duty hoop nets, an open-ended net, at least one large nylon cargo net, at least one canvas tarp, and large plastic dog crates.

Transport kennels will vary depending on the size of the animal. Very small sea lions and seal pups can fit into dog crates that are meant for a cocker spaniel sized dog. Elephant seal pups and sub-adult sea lions can fit into large and jumbo-sized carriers meant for big dog breeds. For larger animals, rescuers should consider investing in a customized aluminum transport carrier. To make the transfer from beach to vehicle easier, some marine mammal experts use carts with large wheels.

Over the years, specialized pinniped capture equipment has evolved to make it easier and safer for the rescuers and the animals. The following details some of these tools and how they are used.

Hoop nets

Pinnipeds that are the size of a springer spaniel or smaller, up to about 60 lb (27 kg), can be netted using a strong-handled hoop net, such as a salmon net. The netting material should be knotless. Once the hoop is over the animal's body, the idea is to manipulate it into the sock of the net. Then, the hoop can be twisted to prevent the animal from escaping.

When the animal is securely within the confines of the net there are a couple of different methods for getting it from the net to the transport carrier.

Small or weak pinnipeds can be manipulated by hand. A dampened towel can be placed over the animal's head, blocking its vision and adding an extra layer of protection for handlers. Wearing heavy-duty thick gloves, one handler will want to restrain the animal's head through the towel, while others peel away the net to expose the animal. At some point the handler will have to switch his or her grasp so that the net can be removed all the way. The animal can then be helped into a carrier.

Another method of getting a small pinniped from a hoop net into a carrier is to swiftly sandwich the hoop of the net up against the opening of the carrier and hold it flush while herding boards are used to encourage the animal inside.

Modified open-ended hoop net

Large, aggressive pinnipeds, such as adult sea lions, can be more safely contained using a modified net designed to reduce the need for actual contact with the animal. The idea is to secure the animal in the bag of the net and then transfer it from the net to a transport crate without having to restrain it by hand (Fig. 62).

When preparing to capture a pinniped using an open-ended hoop net, the fabric within the hoop should be slack, not taut like a tennis racket. If it's stretched too tight the animal can bounce off of it.

When netting the pinniped, the capturer will aim to place the head of the animal in the center of the hoop. The next move will be to pull back on the handle while lowering it to the ground. This is when additional rescuers will move in with herding boards. They will use their boards to encourage the animal over the hoop and into the bag, making sure to also use the boards to protect teammates from being bitten.

This is also the time when the person assigned control of the bag will want to keep the bag stretched out and open, not collapsed around the animal. They want the animal to see a passage so it will be more inclined to move through it, deeper into the sock. Once the animal is at least half way down the sock, two boarders will want to stage on top of the sock, at each end.

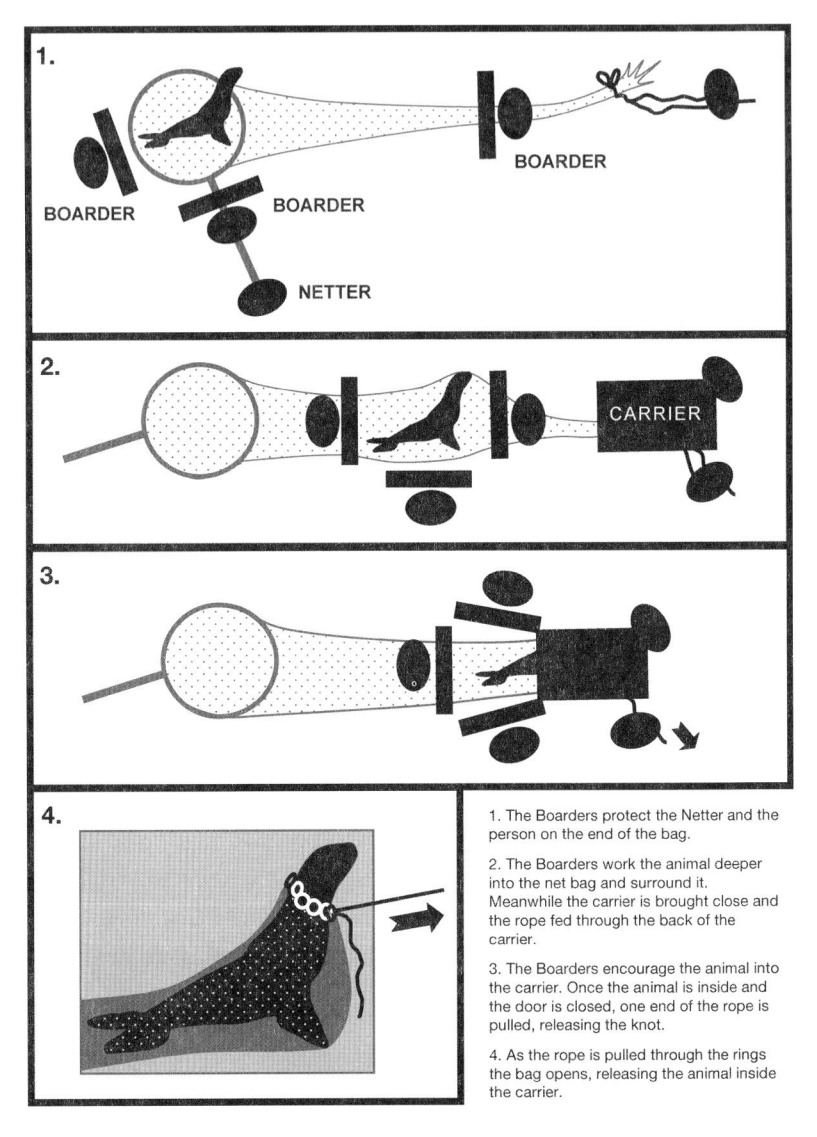

Fig. 62 Boarding a seal or sea lion through a net into a carrier.

1. The Boarders protect the Netter and the person on the end of the bag.

2. The Boarders work the animal deeper into the net bag and surround it. Meanwhile the carrier is brought close and the rope fed through the back of the carrier.

3. The Boarders encourage the animal into the carrier. Once the animal is inside and the door is closed, one end of the rope is pulled, releasing the knot.

4. As the rope is pulled through the rings the bag opens, releasing the animal inside the carrier.

Once the pinniped is securely confined in the bag, the open end of the sock attached to the hoop can be tied off and the hoop and handle removed. If necessary, a large nylon cargo net can be thrown on top of the animal to weight it down.

Typically, once the animal is netted, other team members begin moving the carrier into position near the net. With the carrier in place, the end of the release rope – the rope that is holding the end of the bag closed, needs to be fed through an opening in the back of the carrier. When the team is ready, the boarder stationed on the net, closest to the carrier, will step off the sock, allowing the netted animal to move into the carrier.

Once inside, the cage doors can be closed and locked. The rope holding the sock shut can then be loosened and pulled through, freeing the animal inside the carrier. The remainder of the netting can then be slipped out from under the door.

Wraps, slings, and stretchers

Because sea lions and other otariids are so flexible, mobile, and often aggressive, a hoop net is typically the tool of choice for capture. When containing small or more placid animals, such as elephant seal pups, stretchers and specially made slings can be used.

Using two large wooden or aluminum dowels, rescuers can attach heavy-duty, knotless webbing to create a type of sling for carrying seals. To use the sling, it is draped over the animal with the two handles laid on either side. Rescuers then scoot or roll the seal into the sling as it is tipped laterally and raised.

A similar device, essentially a foldable hoop net, resembles a taco when collapsed over an animal. The frame can be made out of heavy-duty PVC electrical conduit. The net bag should not be so deep that the animal's body touches the ground when lifted.

Wrapping an animal is another method for quick transfer from the shoreline to carrier or transport vehicle. The "burrito wrap," as it has been called, entails folding material, such as canvas or cargo netting, over the animal so that it cannot get free. First, the animal must be convinced onto the large stretch of material using herding boards. The fabric is then folded over the animal, enveloping it (Fig. 63). Using caution and watching out for the animal's head, the bundle can be lifted and transferred into a carrier or vehicle where the material must be removed. Marine mammals overheat easily and must not be kept wrapped in cloth or fabric for more than a few minutes.

The towel wrap

The towel wrap is a method of seizing a smaller pinniped without the use of a net. This requires a large towel, dampened to increase its weight, making it easier to wield with accuracy. The handler should wear thick gloves in preparation for handling the animal.

On approach, a pinniped will typically raise its head, readying to defend itself. Gripping the longest edge of a large towel, close to each corner, the handler

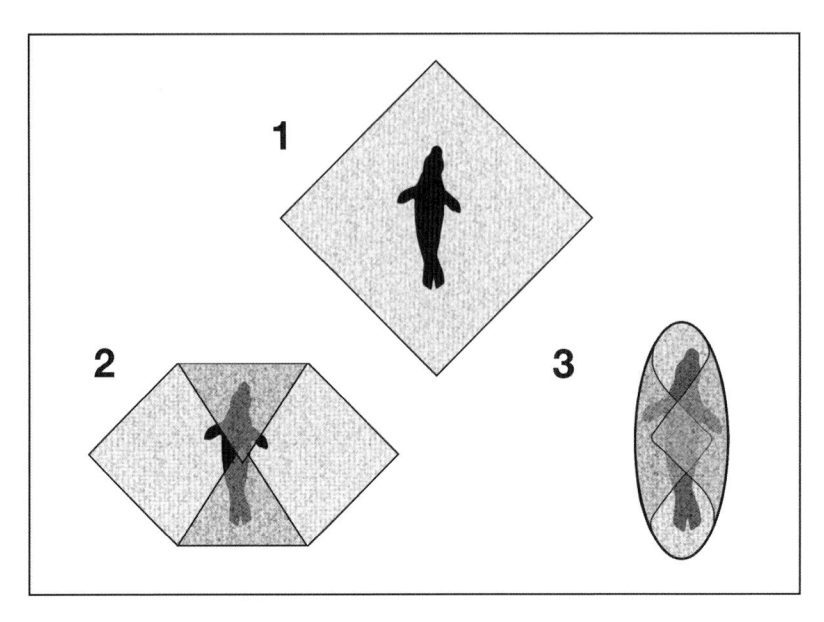

Fig. 63 The burrito wrap can be used to restrain a seal in order to carry it off the beach or load it onto a transport vehicle.

will want to cast the material out and over the head of the animal, much as one would cast a throw net. Without letting go of the fabric, the handler will then cross their hands, wrapping the ends of the towel around the animal's head. When the animal's head is wrapped, the handler, in one fluid movement, will want to gently but firmly grip behind the animal's skull and drop into a classic pinniped restraint position, on their knees. The animal can then be manipulated into a carrier and the towel removed.

Flat webbing cargo net

Cargo nets that are made of flat nylon webbing are typically soft and pliable, yet extremely strong and durable. They can be used to help weigh down a large, aggressive pinniped that has been captured in a hoop net, or used as a type of sling to carry seal pups.

Elephant seal pups, for example, have a tendency to roll when they are manhandled. This natural behavior can be used to help rescuers roll a pup into a cargo net sling. With its head being controlled, the animal can then be carried and placed into a carrier and the cargo net removed from under it.

The floating net

A floating net can be constructed out of PVC. Netting is attached to form a large, floating pen, or net. A drawstring is used to close the net once an animal has entered it. A second line is secured, not to the PVC frame, but to the net bag to prevent a large animal from swimming off with it. These nets are used to capture animals that are close to the water's edge, such as a sea lion at the end of a dock.

The method requires that one rescuer, preferably wearing a wet suit and fins, enters the water without being noticed by the target animal. He or she will swim, floating the net in front of them, aiming to slip the net up next to the sleeping animal without being seen or heard.

When the floating net is in place, one of the rescuers will approach the animal on land, holding a hoop net in front of them in case the animal charges. If all goes as anticipated, the animal will plunge into the floating pen.

Once the animal enters the trap, the top must be drawn closed. The PVC frame can be twisted or flipped over to help keep the animal from escaping.

While the rescue team readies to lift the netted animal from the water, they must ensure that the animal is able to breathe.

Physical restraint of seals and sea lions

When manually restraining a small to medium-sized pinniped, one must have control of its head. Usually this is achieved through a steady hold from behind – the handler cupping the animal's neck, with their thumbs on the back of the skull and their fingers against the angle of the lower mandibles (Fig. 64). If the seal is larger than a cocker spaniel, its body may need to be controlled.

A medium-sized seal or sea lion that is smaller than a German shepherd dog can be physically restrained by being straddled. The handler will want to first gain control of the animal's head, then straddle its shoulder area taking care not to "sit" on it. The flippers should be gently tucked back in normal position against the animal's body and held in place by the handler's inner thighs (Fig. 65). Larger and more boisterous animals will require additional handlers.

Confinement and transport of pinnipeds

Transport containers should be large enough to give an animal room to stretch out and raise its head. Most plastic animal carriers, such as those approved for air travel, will be suitable for most juveniles and small adults. Small seal pups can be confined in plastic animal carriers or plastic tubs, nearly anything, so long

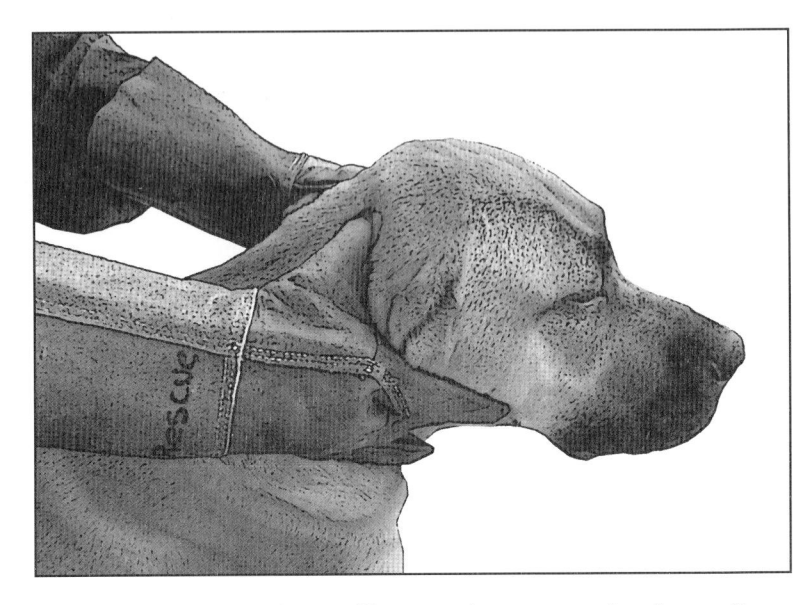

Fig. 64 Handlers use a dog to illustrate the proper grip of a seal's or sea lion's head.

it affords a solid lid and there are enough air holes for adequate ventilation. Large adult pinnipeds may need specialized, purpose-built containers (Fig. 66). For transport that lasts over four hours, animals should be housed in larger confines that are half their length in width or greater, allowing them to move freely. One solution is to have them travel in the rear compartment of a vehicle.

Vans, SUVs, even hatchbacks, can be modified to safely transport some species. Plywood can be used to block access to the front of the vehicle. Heavy-duty waterproof material can be used to line the compartment, using a PVC frame to hold up the sides. Anti-fatigue mats covered with damp towels can provide traction and will help trap excrement. A rear air conditioner may be necessary to keep the area cool. A number of animals can travel together nicely this way, especially elephant seal pups.

Marine mammals can overheat quickly, especially active juvenile and adult pinnipeds in good condition. These animals should be given damp towels to lie on. During transit they can be sprayed with water regularly, even their eyes, to keep them moist. As a rule, animals should be checked on at least every two hours. On journeys that last over four hours, animals should be provided with a rest period of at least 45 minutes.

Also, on long journeys or situations where spraying the animal with water is impossible, trays of ice cubes can be fixed to the roof of the carrier – as they melt, cold water will drip down on the animal. Cold packs can also be placed

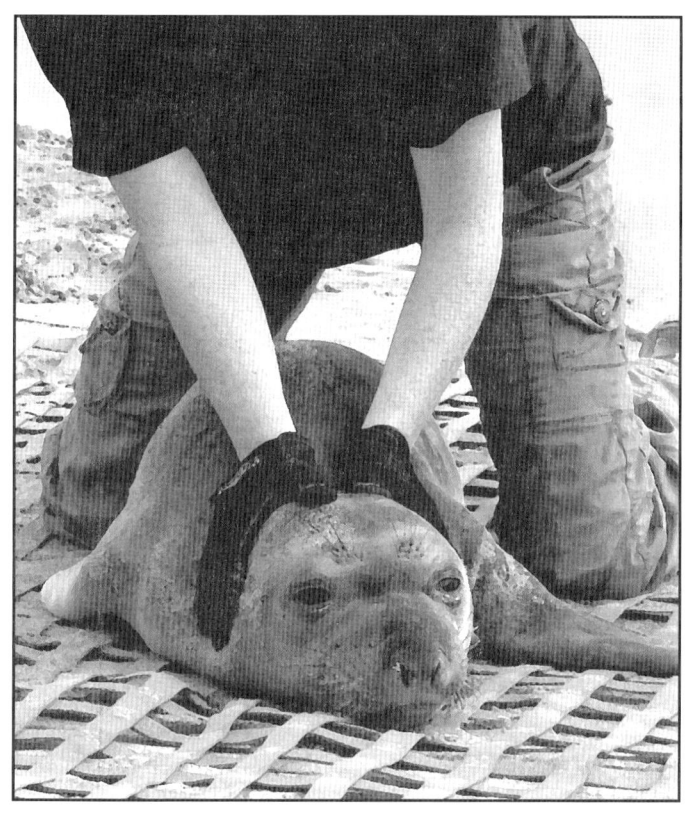

Fig. 65 While keeping control of the head, the handler will straddle the animal, folding the flippers back and keeping them tucked against their legs.

under the damp towels. If the transportation is in an enclosed environment, the ambient temperature should be kept close to 50° F (10° C). For emaciated animals or newborn pups, however, this may be too cool.

In general, very young seal pups will not be prone to overheating, just the contrary. In some cases, they may even need supplemental heat. Young pups, or emaciated animals, should be transported in an enclosed, well ventilated but temperature-controlled environment. Animals should be provided with dry blankets or towels to lie on. If a pup is easy to handle safely, taking its temperature before travel will help in determining its needs. If transport is to exceed two hours, it may be wise to check its temperature during the journey.

In preparation for travel, containers should be firmly secured in place and not allowed to slide or move about during travel. Transporters should have all necessary documentation for transporting marine mammals, including a vehicle placard indicating there are live wild animals aboard. This sign should include the

Fig. 66 An example of a lightweight specialized cage used to transport large pinnipeds.

phone number to call in case of emergency. As a policy, the doors of the transport vehicle should be kept locked at all times.

Cetaceans

Different countries have differing policies regulating the handling and rescue of cetaceans. Rescuers should be aware of these rules before assisting an animal.

The following basic handling suggestions are provided for those involved in assisting beached whales, dolphins, and porpoises.

Initially, rescuers will want to tend to the stranded animal and its immediate needs. Before making additional rescue plans, the focus should be on getting the animal stabilized and protected from the elements. If rescuers can safely hold the cetacean afloat, this will be less stressful for the animal than bringing it ashore. When floating the animal is not an option, rescuers should consider stabilizing it on shore, out of the surf zone.

On land, a cetacean must be supported in an upright position without too much pressure being applied to its tail or pectoral fins. If possible, the animal should be placed on foam padding. Rescuers may need to dig trenches to accommodate its fluke and front flippers.

Most casualties will need to be kept wet and protected from the sun, and overheating. When possible, stranded cetaceans should be covered with sheets or towels, which should be continually sprayed or doused with cool water. Extremely cold water or ice must never be used on flukes or fins.

It is extremely important that the animal's blowhole is kept clean and unobstructed, and that water and sand are prevented from entering it. The blowhole's margins can be protected by smearing them with lubricating jelly or zinc oxide cream.

To protect an animal's skin from sunburn, rescuers should build a protective shade. If necessary, a windbreak should be built to protect very young or emaciated animals from cold winds. To keep these more vulnerable animals from getting too cold, rescuers can soak the protective sheets in mineral oil rather than with water.

Unlike humans, cetaceans are conscious breathers, meaning they have to make a conscious decision to breathe, or not. A rescued animal's rate of breathing should be monitored closely. As a guide, the normal breathing rate for small cetaceans is roughly two to five breaths per minute. Medium-sized cetaceans, such as pilot whales, take in one breath per minute normally, whereas larger whales can have respiration rates as low as one breath per 20 minutes. If a cetacean becomes too stressed, it may start holding its breath, or delaying inhalation. Research has suggested that some cetaceans will respond well to individual human contact. A calm, soothing voice and gentle touch have been known to calm some cetaceans.

12 Basic wildlife first aid and stabilization

From the time an animal is reported to the time it reaches definitive care, responders have the opportunity to increase its chances of surviving through proper handling, appropriate housing, and the provision of first aid.

The administration of first aid can begin early on, during the Call Taker's first conversation with the finder. Through effective questioning they might, for example, establish that a baby bird is suffering from hypothermia (lowered body temperature), a life-threatening condition. By guiding the finder through safe methods of providing warmth they greatly increase the chance that the baby bird will survive.

Once on scene, an animal's condition should be assessed thoroughly to determine its immediate needs. If the animal has not yet been captured and confined, and when injuries are not readily apparent, rescuers should try to observe the animal from afar, out of sight if possible. Fear can override pain, allowing animals to hide their weaknesses. In the presence of human predators, certain animals may stop pain guarding, or limping; birds may react by raising an otherwise drooping wing.

When observing a bird with a potential wing problem, pay close attention to exactly how the wings are held (Fig. 67). They are normally held tight against the bird's body, not limp, or dragging. A drooping wing can be caused by injury to the bones or muscle, or nerve paralysis, or, when both wings are equally affected, it might be due to muscle weakness or exhaustion. If the primary feathers are trailing on the ground, this can indicate an injury to the radius, ulna, or digits. If the wing is noticeably drooping but the primaries are held up off the ground, the injury might be to the elbow or humerus. Shoulder or coracoid injuries can cause the wing to rotate slightly so that, while the wing might droop slightly, the primary feathers might appear tipped up in relation to the uninjured wing.

While observing an animal, decide what its immediate needs are and how you are going to provide for these needs with the least amount of handling. If the animal requires immediate medical aid, have all the necessary supplies and equipment assembled ahead of time, even before it is captured and confined.

Once in hand, most animals will be experiencing tremendous stress. Additional handling may endanger their lives. This is no time to practice or to show off, but a time for rescuers to seriously evaluate the benefits of added time and handling.

Wildlife Search and Rescue: A Guide for First Responders, First Edition. Rebecca Dmytryk.
© 2012 John Wiley & Sons, Ltd. Published 2012 by John Wiley & Sons, Ltd.

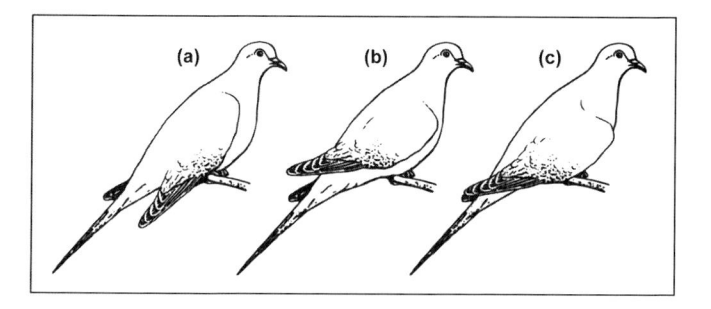

Fig. 67 (a) The primary feathers are hanging low to the ground, indicating possible trauma to the radius, ulna, or digits. (b) When primary feathers are offset – held higher than the normal side – this can indicate a shoulder or coracoid injury. (c) When the wing droops but the primaries are folded normally, this can indicate injury to the elbow or humerus.

Administering first aid in the field might simply delay proper treatment at a wildlife facility; however, if the efforts would prevent an animal's condition from worsening, it would then be reasonable for rescuers to administer aid.

Performing a cursory physical examination

Most wild animals will perceive any handling by humans as an attack. Unless it is absolutely deemed a necessity, the physical exam should be performed by a wildlife rehabilitator, not by rescuers in the field. If rescue personnel must perform the exam it should be completed within thirty to sixty seconds.

To perform a physical examination, begin at the head. The eyes of the animal should be symmetrical and centered in the orbit. They should be clear, glossy, and bright, not sunken or "deflated." Note any rhythmic twitching, or nystagmus. Check the pupils for signs of abnormal dilation. Using a penlight, see how the pupils react to its brightness. Look for signs of bleeding or welling of blood in the eyes. Occasionally, animals will be so injured or incapacitated that they cannot close their eyes or blink. The eyes can be flushed with sterile saline solution and protected with a non-steroidal eye lubricant.

When looking at the mouth of any animal, rescuers must be extremely careful, as many species can deliver painful and damaging bites. When examining the mouth, note its overall wetness – is the saliva stringy or sticky or does the mouth appear dry? The color of the gums, tongue, and roof of the mouth should be noted. Are there nodules or growths visible? Is there any blood present? Glance at the dentition and note any broken or missing teeth. Rodents have teeth that continually grow, causing problems when they get too long.

When looking at the ears, check for any bleeding or any type of discharge, or the presence of fly eggs or maggots. When examining the nares, or nostrils, note any discharge and what color it is. If there are sounds during breathing, do they emanate from the nasal area, or from deeper in the chest?

Rescuers should note the animal's body condition (thin or emaciated). Land mammals should feel well muscled. When evaluating birds, the breastbone, or keel, should have muscular padding on either side; it should not be "sharp" or concave. Marine mammals should be well-fleshed and "rounded".

If the animal is in a critical state, address life-threatening conditions first. Start by evaluating the animal's vital life-support systems – its breathing and blood circulation.

Establishing whether or not an animal is breathing is usually simple – by observing the movement of the rib cage. In reptiles and unconscious birds, it may require a look inside the animal's mouth for the opening and closing of the windpipe.

In evaluating the animal's breathing, note the rate of respiration – how many breaths the animal takes in one minute. A high respiration rate, especially if it is accompanied by open mouth breathing, or panting, can be a sign of overheating. Also, when checking the animal's respiration, listen for sounds that might indicate blood, mucus, or fluid in the trachea, or "sucking" sounds that might indicate a puncture wound. Attempts should be made to seal any puncture wounds penetrating the airway. One method would be to apply an airtight bandage using plastic wrap, gauze, and tape.

If the animal is not breathing, rescuers must act quickly. If an animal is deprived of oxygen for too long, between three to five minutes, it can suffer irreversible brain damage. Sometimes, repositioning the head can help to open the airway. Check the airway for an obstruction, like or a foreign object, or blood. Mucus or blood can be swabbed or suctioned away. If something is blocking the windpipe, attempt to remove it, using forceps if appropriate, and take care not to push it deeper. If the airway is clear, and the animal is not breathing, artificial respiration can be administered with an ambubag or through an endotracheal tube, which only a veterinarian or properly trained wildlife medic should attempt.

Once breathing is established, the animal's circulatory system can be evaluated. Establishing the heart rate is common practice, though for wildlife rescuers in the field this information will be of little consequence. However, for times when it is important, the heart rate can be established using a stethoscope. A stethoscope has two heads, a bell and a diaphragm. The bell head picks up lower frequency sounds more clearly. It can be held against the animal's chest wall over the heart. Count the number of heartbeats in 15 seconds and then multiply that number by four to calculate the number of beats per minute (b.p.m.). The pulse rate is different than the heart rate. The pulse rate and strength can be taken and assessed at various locations on the body and the best locations to do this vary by species. The pulse indicates how well the heart is pumping blood through the arteries. Ideally, the pulse rate will be the same as the heart rate.

Another way of judging the health of the circulatory system is by the color of an animal's mucosa, or mucous membranes. The color of these moist layers of tissue, which are typically bright pink, is important because it indicates good blood flow and plenty of oxygen in the bloodstream.

Gums can also provide information on the health of an animal's circulatory system through what is called capillary refill time (CRT). Pressing down on gum tissue pinches closed the small blood vessels, or capillaries, causing the tissue to turn pale. Once the pressure is released, the capillaries refill with blood, and color returns. Normal CRT is between one and two seconds. Slow refill time indicates circulatory problems.

Bleeding

If the animal is bleeding, a gauze pad can be used to apply direct pressure. If the blood soaks through, additional layers of bandaging should be applied. With luck, the blood will clot and the wound will stop bleeding. It is important, though, that the area not be wiped or dabbed, or the wound may start bleeding again if the clot is dislodged. Handle bleeding animals only when absolutely necessary – handling increases blood pressure and can impede clotting.

Birds bleeding from the mouth should be handled as absolutely little as possible and should not be turned onto their backs. Put these birds in a comfortable position in a dark box and transport them to a care center right away.

Sometimes, a bird may be found bleeding from a feather, called a blood feather. These are new feathers that are still attached to a blood supply. A broken and bleeding blood feather can be grasped firmly at its base and pulled straight out in the direction of its growth. Pressure can then be applied to stop the bleeding. Cornstarch makes an excellent non-toxic clotting agent to help stop bleeding from blood feathers.

When any animal loses too much blood, it can go into what is called hypovolemic shock. This is when the volume of blood circulating in the body is too low due to hemorrhaging, external bleeding, or fluid loss causing severe dehydration. These animals need immediate veterinary attention.

Dehydration

Dehydration can be a life-threatening condition. Most wildlife casualties presenting with some form of distress will be suffering from some level of dehydration.

The degree of dehydration can often be determined by the animal's appearance. Its eyes should be glossy and alert, not sunken or dull; its mouth should appear wet and the saliva watery, not stringy and sticky. Its mucous membranes should be moist, not tacky or dry.

Animals that are severely dehydrated or in shock will tend to be hypothermic (cold) due to widespread vasoconstriction – the body's attempt to keep blood circulating to the brain and heart but away from the skin. Treatment begins by providing warmth and replacing fluids. Animals should be warmed to approximately normal body temperature before fluids are administered. The only time it is reasonable for a "cold" animal to receive fluids is if they are delivered intravenously.

Fluid therapy

Once an animal has been warmed to normal body temperature, fluid therapy is used to treat shock by helping increase blood volume and blood pressure. It is one of the most basic supportive treatments – one that can make a difference between life and death. However, there are many risks associated with giving fluids to an animal. For example, oral fluids can be aspirated (inhaled), potentially causing pneumonia or even drowning, and overzealous administration of subcutaneous or intravenous fluids may cause over-dilution of the blood in anemic animals or fluid overload in any animal whose organs are not functioning well enough to remove excess fluid from the body. Prompt transport to a care center should take precedence over starting fluid therapy in many cases, but if transport to definitive care is going to be delayed several hours or overnight, fluid therapy is essential. Some animals may regurgitate during transport and be at risk of aspiration; hence caution is advised when orally hydrating animals prior to transportation.

Water is the principle component of fluid therapy. In addition to water, all animals' bodies require minerals, known as *electrolytes*, when ionized, or *salts*, when bound to another chemical compound. Essential electrolytes include comparatively large amounts of sodium, potassium, and chloride, and smaller amounts of others such as calcium and magnesium. Hormones in the body regulate the absorption and excretion of these electrolytes to achieve homeostasis, or balance.

Tonicity refers to the amount of dissolved substances (electrolytes, sugars, others) in a fluid. *Hypotonic* fluids have fewer dissolved substances than a typical animal's tissues, *hypertonic* fluids have more. *Isotonic* fluids have concentrations similar to animal tissues, and thus are ideal for replacement of both fluids and electrolytes in dehydrated animals. Common preparations of sterile isotonic fluids for injection include lactated Ringer's solution (LRS), Normosol, 0.9% NaCl, and

0.45% NaCl/2.5% dextrose. These fluids may also be given orally. Sterile fluids with more than 2.5% dextrose may not be given subcutaneously, as higher concentrations of sugar may result in the fluids not being absorbed; 5% dextrose solution is a common fluid intended for injection that is not appropriate for subcutaneous delivery. If using an unfamiliar fluid, check the label for a number describing the osmolarity for the solution. If this number is outside the range of 270–310 mOsmol/l, do not use it without guidance from a medical professional. This number shows the total concentration of solutes in the fluid, and thus its appropriateness for various uses.

There are many types of fluids available for fluid therapy, and several routes of administration, such as oral, subcutaneous (injected under the skin), and intravenous (injected directly into a vein). Principle routes for reptiles also include intracoelomically (injected into the abdomen) and by soaking the animal to allow it to absorb water cloacally. Untrained caregivers should not attempt injectable methods of fluid administration, and animals that have difficulty swallowing, have been vomiting, are having seizures, or are unable to hold their head upright should not be given oral fluids.

Oral electrolytes solutions include those intended for rehydrating human children, sports drinks, and homemade electrolyte solutions such as 9 grams of table salt (sodium chloride) in one liter of water. These solutions may have bacterial contamination because they are not prepared in a sterile manner, and thus may not be given to an animal by any other route than oral. When choosing an oral rehydrating solution for wild animals, avoid choosing fluids with large amounts of sugar. Carnivorous animals may become hyperglycemic when given fluids containing sugars. This is less of a consideration for omnivorous or herbivorous species.

Fluids intended to be injected into an animal are manufactured in a sterile manner and come in a plastic bag or bottle that has a stopper through which a sterile needle is inserted to remove fluids for use. Once these containers are opened to the air they are no longer sterile and may be used as oral fluids only. Only clean sterile needles may be used to puncture a fluid containers stopper. Sterile needles are no longer sterile if they are uncapped and touch anything such as your hand or a table. If a used needle is punctured into a container, the fluids are no longer sterile.

The more severe the dehydration, the more invasive is the ideal method of rehydration. Animals with simple dehydration, such as due to being trapped without access to water, often have not lost substantial amounts of electrolytes and may be successfully rehydrated with plain water given orally. Animals that are very young, emaciated, sick, have wounds, or have lost blood, need replacement of electrolytes as well as water. The method used to rehydrate these animals is dependent on the patient's species and health status, the availability of supplies, and the skill of the caregiver. In the field, the most practical methods for delivery will be orally or through subcutaneous injection.

In all cases it is important that animals undergoing fluid therapy be monitored for their reaction and signs of improvement. As an animal becomes hydrated its eyes will become brighter and it will be more alert.

Unless an animal is suffering from hyperthermia (overheating), fluids should be warmed to the animal's normal body temperature before being administered. The amount given is based on the calculated amount an animal needs daily under normal conditions and adjusting for ongoing losses (diarrhea, vomiting, burns), plus restoring fluid deficits. This total amount is typically administered over a 72-hour period with half the fluid administered in the first 24 hours, with the remainder delivered over the subsequent 48 hours. Most mammals require 80–100 ml/kg of fluid per day for maintenance; birds require 100 ml/kg per day; reptiles require 25 ml/kg per day.

An animal's fluid deficit can be calculated by multiplying its level of dehydration by its body weight: fluid deficit (ml) = % dehydration × body weight (g). It is reasonable to presume 10% dehydration in most of the injured or ill animals encountered, especially in cases where there has been blood loss or fluid loss from diarrhea or vomiting, or if the animal is exhibiting signs of shock.

The total amount of fluids is not to be administered all at once but spread throughout the day. It can be divided and administered at regular intervals depending on the amount the animal can handle. Stomach capacity in most species of mammals and birds is approximately 50 ml/kg, or 5% body weight. Initial oral hydration volume must not exceed this amount or the animal may regurgitate and risk aspiration. For example, a 1000 g hawk could be expected to have a stomach capacity of 50 ml. A prudently conservative amount for initial oral hydration for weak animals would be much less than this amount, 10–25 ml/kg. If this amount is well tolerated, the volume may be raised at subsequent administrations. The risks and stress of handling may need to be weighed against the benefits of more frequent handling. Adult animals are likely to be more stressed by frequent handlings than juveniles, and may be more dangerous to caregivers.

Oral fluid administration (mammals)

Some small to medium-sized mammals can be offered fluids through a syringe filled with the correctly calculated dosage. If possible, this should be done in a quiet setting as mammals can become squirmy and lose focus if there are too many distractions.

The tip of the syringe or tube can be introduced at the front or side of the mouth, gently and slowly offering a drop of the solution so the animal can taste it. If the animal reacts by drinking, the caregiver should try to provide an uninterrupted flow of the solution, keeping steady pace with its consumption. Delivery of fluid in birds is quite different.

Oral fluid administration (birds)

Providing fluids to birds orally is the least stressful and most practical method, but it takes skill and training to do it properly. This is especially true when giving oral fluids to small birds. Because of where the windpipe is located – at the back of the tongue, it is very easy to "drown" them. However, with the right amount of training, it is possible to safely administer oral fluids to small birds. A 1 ml syringe with a plastic cannula tip is perfect for many species, such as finches, goldfinches, and swallows, and they are easy to clean. A plain 1 or 3 ml syringe is ideal for larger chicks, such as young robins, and corvids.

For larger birds, a long feeding tube can be used to deliver fluids directly into the stomach or crop (in species that have a crop). This is called gavage, or tube feeding. The chief concern when administering oral fluids is making sure no fluid gets into the windpipe, from either incorrect delivery or aspiration of regurgitated fluids.

The glottis is the opening to a bird's windpipe, or trachea. It is located at the base of the tongue (Fig. 68). Just above it, on the roof of the mouth, is a slit, called the choana, which connects to the nares (nostrils). In most birds, when their beak is shut, the glottis fits snugly into the choanal slit forming a closed connection.

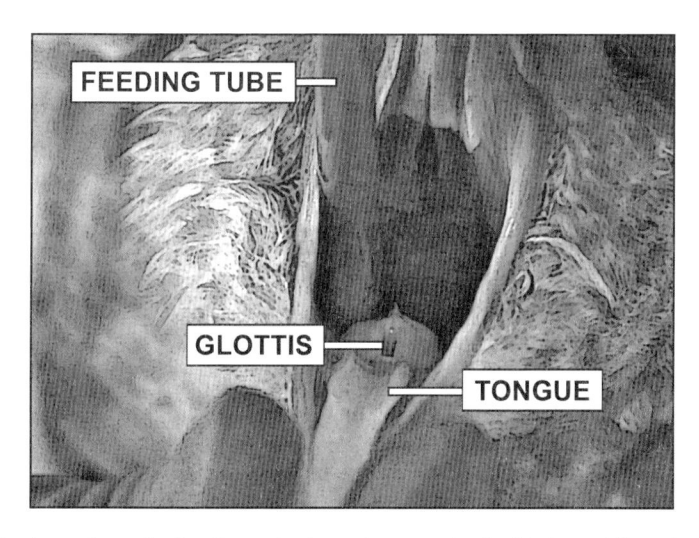

Fig. 68 Insertion of a feeding tube into the mouth of a bird, avoiding the glottis.

In large and medium-sized birds, the glottis is fairly easy to identify and avoid. In smaller birds, it is usually very difficult to see.

Because of the delicate nature of the procedure, gavage should only be performed by those with experience, or under direct supervision of someone who is skilled at tube feeding. Following certain steps and protocol will lessen the chance that mistakes will happen.

One such rule is to avoid excessive handling of the bird before and after administering fluids. Caregivers should prepare the fluids and have them drawn and ready to administer before picking up the bird. If an avian patient must have its blood drawn, or be weighed, or have bandages changed, fluids should be administered last to reduce the chance of regurgitation.

For avian species with long necks such as gulls or ducks, fluids are delivered through a French catheter attached to a syringe. Customarily, fluids are drawn up through the tube. To clear air bubbles, hold the syringe pointing up and depress the plunger to express the air. To warm fluids prior to administration, the syringe with catheter attached can be submerged in a pitcher of warm water. In the field, fluids can be kept warm using a commercial beverage or baby bottle warmer. Another way to keep fluids warm in a vehicle is to set the syringe on the warm defrost vents on the dashboard.

To perform gavage, caregivers should work as a team; one person restrains the bird while the other performs the gavage. When performing gavage on a duck or gull, for example, the handler will want to hold the bird at hip-level, gently braced against their body with a sheet or towel over the bird's head to reduce stress. As a rule, the handler maintains control of the bird's head until the "tuber" requests control of it. If the "tuber" is right-handed, the bird is usually held on the handler's right side, and on the left side for left-handed tubers.

If performing gavage on a raptor, the handler will use proper restraint techniques, depending on the species, and make sure the bird is held at the right height for the tuber. A small, lightweight towel can be used to restrain songbirds for this procedure, as it can help keep their wings and legs immobilized.

Customarily, the tuber will approach the handler and take control of the bird's head, communicating with the handler as needed. The tuber will usually have to extend the bird's neck. Unless instructed otherwise, the handler will want to hold the bird in place, bracing against the tuber's gentle pull. The tuber will softly open the bird's bill and insert the appropriately sized French catheter, or tube, aiming for the back of the oral cavity and the right side of the bird's throat.

As the catheter is gently inserted, watch and feel for signs that it is passing down the esophagus. If any resistance is felt pull back slightly and try again, or try adjusting the angle of the neck. Check to make sure the tube is not in the trachea. This may be accomplished in two ways. In large birds, such as gulls, it is sometimes possible to peer inside the bird's mouth. If the bird's mouth will not open wide enough to see the glottis, feel the front of the bird's throat for two "tubes" – the trachea and the catheter in the esophagus. Never force the mouth open further than it will open with gentle pressure or the jaw may be damaged.

Once the catheter has been inserted far enough, the tuber will slowly and steadily begin to deliver the warm fluids. They will want to watch for signs of the fluid welling up in the throat, such as a bulge forming at the base of the neck. If the bird begins to regurgitate the catheter should be removed and the bird given the freedom to shake its head and expel fluid away from its airway. A bird's bill must never be held shut when it is regurgitating. Whenever the catheter is being removed, it should be pinched or folded to prevent any liquid from dripping into the trachea.

After the fluids have been administered the bird should be placed back into its container promptly. Keep the bird's head higher than its body and allow the bird to steady itself before letting go. These small measures will help avoid regurgitation.

When administering oral fluids to pelicans, the technique is slightly different. The bird can be covered in a sheet and left standing, restrained by the handler. The handler can either straddle the bird, using their legs to keep the bird's wings from flapping, or kneel, holding the bird against their midsection (Fig. 69). If they can,

Fig. 69 A young pelican is restrained by a carer who is about to administer oral fluids.

the handler will help keep the bird restrained, its bill open, and its neck extended. Customarily the person administering the fluids will stand in front of the pelican and use both hands to deliver the fluids.

The pelican's trachea is very obvious and therefore easily avoided. It is situated about two-thirds of the way down the pouch. The small fleshy projectile is the pelican's tongue. Pelicans usually require a good amount of fluid, which can be simpler to give without the catheter – just straight from the syringe. This also eliminates the chance of the catheter falling off and getting lost in the pelican's throat. The fluid should be delivered down the throat, well beyond the trachea.

After the fluid has been administered, the handler should continue to hold the pelican's neck extended while the base of the neck is massaged, gently pushing the fluid deeper into the gut.

As the animal is released into its enclosure, care must be taken that it does not fall forward, as this can bring on regurgitation. Stress will also cause them to regurgitate. If administering fluids to numerous pelicans in a small enclosure box, a sheet can be used to block the pelicans' view of the handlers while also helping to separate out those that require treatment.

Subcutaneous injections

Subcutaneous delivery of fluids is often the most practical method. Fluid is injected into the space between the skin and muscles and is absorbed into the bloodstream over time. However, cold animals or those with poor circulatory function due to dehydration or blood loss will have poor absorption of fluids from this route. The main advantage over oral fluids is the lack of concern for regurgitation and aspiration.

For most mammals, the preferred site for delivery is the area between the shoulder blades and all along the back on either side of the spine, wherever the skin is stretchiest. In birds, the most commonly used site is the thigh region where the skin is quite stretchy due to the movement of the knee. It is safest to give fluids to birds in this area by injecting on the outside of the thigh near the knee. Never direct a fluid administration needle towards the abdomen. There are abdominal airsacs which may be penetrated and there is a risk of drowning the bird if these airsacs are injected.

Birds may also be given fluids under the skin of the back over the ribcage, but in many species this area cannot accommodate much volume. Due to the presence of air sacs, the dorsal neck region should be avoided. Never give subcutaneous fluids in the patagium (wing webbing). Due to having air sacs on the outside of the ribcage, pelicans should not receive subcutaneous injections.

Always have the hand operating the syringe ready to depress the plunger without an adjustment of grip, as movement to the needle may tear the skin and cause the fluids to leak out the hole.

A single subcutaneous fluid injection site may accommodate approximately 50 ml/kg (5% body weight) in mammals and birds, although this may vary depending on the species and the location. Fluids must be warmed to normal body temperature before administration, either before or after being drawn into the syringe. The needle size must also be considered. For very small birds that weigh under 100 grams, a 25–27 gauge needle can be used, and for those weighing 100–500 grams, a 23 gauge needle works well. For large birds such as geese, a 22 or 23 gauge needle may be used. Mammals have tougher, thicker skin and require larger needle sizes per body weight: 23–25 gauge works well for infants, 20 gauge for adults.

Before drawing the fluids, the puncture stopper on the fluid bag should be disinfected with an alcohol swab. After drawing up the appropriate amount of fluid the syringe can be held upright and flicked with a finger to get the air bubbles to rise; air can then be expelled by depressing the plunger slowly until a drop of fluid appears. The injection site can be disinfected with an alcohol swab as needed. To administer the fluid to a mammal, pull up gently on the skin, forming a "tent." The needle can then be inserted, bevel up, parallel to the animal's body, taking care not to puncture through the other side. The fluid should be delivered slowly, making sure no fluid is not leaking. A small rounded "pillow" of fluid should begin to form. This can be referred to as a bolus. If the fluid is leaking a lot, choose a new location. Try to avoid making several puncture holes in the same location as leaking problems can detract from the benefit of fluid administration.

Amphibians are able to absorb water through their skin. Amphibians suffering from dehydration can be placed in a shallow bath of amphibian Ringer's, at an isotonic concentration. The solution can be made by mixing one liter of distilled water with 6.6 grams of NaCl, 0.15 g KCl, 0.15 g $CaCl_2$, 0.2 g $NaHCO_3$. If amphibian Ringer's is not available, an emergency rehydration solution can be made by mixing 1 part Normosol with 1 part 5% dextrose in water (D5W). The temperature of the bath should be a few degrees below the amphibian's preferred optimal temperature zone (POTZ). In severe cases, artificial slime, such as Shieldex, can be added to the bath after 12 hours.

Reptiles may be started on a hydration regime at 25 ml/kg per day for maintenance plus the estimated deficit based on the animal's level of dehydration. This is typically administered subcutaneously or intracoelomically over 48–72 hours along with slow warming. Reptiles should not receive fluids that contain lactate. Instead, a rehydration solution can be made by mixing 1 part Normosol to 2 parts D5W.

Treating hypothermia

Hypothermia is a life-threatening condition. If a bird appears lethargic and fluffed up or if its feet feel cold to the touch it is probably chilled. Mammals will also feel cold to the touch and will tend to curl up in a fetal position. North American birds and placental mammals have normal body temperatures warmer than humans, so they should always feel warm to the touch of a warm human hand. Marsupials tend to have body temperatures cooler than humans.

Simply being confined and sheltered from the elements may allow some animals to reach a more normal core body temperature. However, this will not be enough to warm neonatal animals that are chilled, or animals suffering from shock. They require heat that will warm their surroundings, not just a portion of their body – ideally, an incubator or heat lamp. If neither is available, there are a few options for providing an animal with warmth.

As a rule, when you provide supplemental heat, animals must be able to move off or away from a heat source as needed and they must never come into direct contact with a heated surface that can cause injury. Animals must be monitored closely for overheating.

An electric heating pad may be an obvious choice for supplying warmth to a cold animal. Heating pads can be placed under or around a portion of a carrier, making sure to allow the animal some room to move away from or off the heat. For quick, concentrated heat, a heating pad can be shaped into a cone, setting a small animal in the center (Fig. 70). The temperature setting of the heating pad will depend on its distance from the animal and the thickness of layers in between. Typically, heating pads are set to low or medium. Regardless, caregivers must continually monitor the temperature and the animal's reaction to it.

In the absence of an electric heating pad, a vessel filled with hot water, wrapped in a lightweight towel, will give off warmth for almost an hour. Glass will tend to hold heat longer than plastic. The disadvantage is that as the water cools it will start to draw heat away from the animal.

Another method is to microwave a wet washcloth and place it in a Ziploc bag. As with any other heat source, it must not come into direct contact with the animal. The bag can be wrapped in a lightweight dish towel, and placed under or around the animal. This type of heat source will give off warmth for about 25 minutes.

Heated dry rice in a fabric bag can produce warmth for up to an hour, depending on what it is contained in. Dry rice can be heated in a conventional oven or microwave, just until it is hot to the touch. In a microwave oven, this takes about two minutes for two cups of dry rice. Once it is heated, the rice can

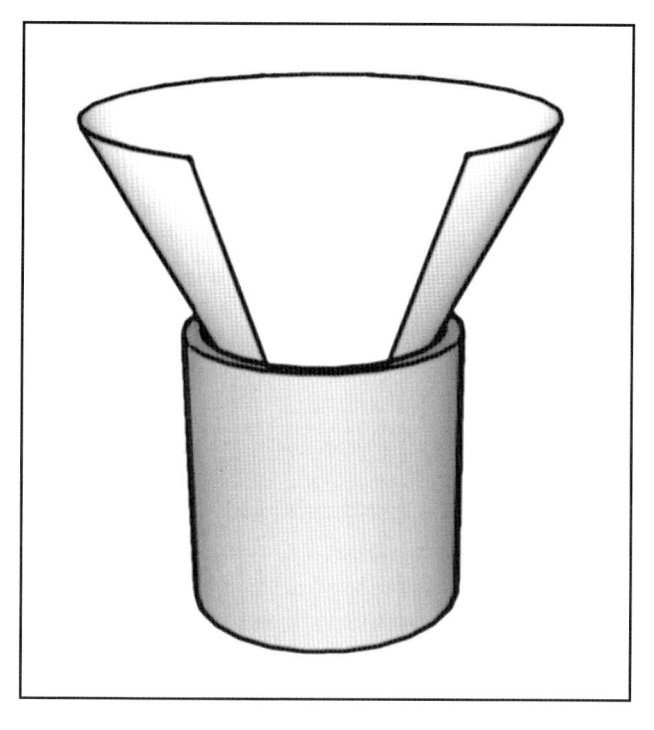

Fig. 70 A heating pad shaped into a cone and set in a container.

be carefully poured into a fabric sack, such as a sock, or into a glass or plastic container. A cotton sock will keep warm for about 20–30 minutes, while something like a plastic storage container will keep warm longer, up to about 45 minutes.

Small mammals can also be warmed by placing them in a plastic leak-proof bag (double-bagged) and floated atop a tub of warm water. The bag must not be sealed but held in place as the animal warms up. This creates a type of mini-incubator.

In primitive situations, dense, igneous stones can be warmed in an oven or a fire. A warmed rock wrapped in a towel can produce heat for up to an hour. If available, pea gravel can be heated in an oven and then pored into a fabric sack or container, similarly to how the uncooked rice is used.

Treating hyperthermia

The opposite of hypothermia is hyperthermia, when an animal becomes dangerously overheated. Environmental factors that can lead to hyperthermia include high temperatures, humidity, and lack of airflow. Wild animals can also overheat under the strain of capture, handling, and restraint, or in their carriers. Animals that are constantly trying to escape will build up heat quickly, especially those with thick fur or dense feathers. Juvenile animals found on paved surfaces such as sidewalks during hot conditions may be dangerously overheated upon rescue.

Hyperthermia is a life-threatening condition. Clinical signs include rapid pulse and increased respiration. Mammals may pant and begin to drool. The color of the inside of their ears may turn pink. An overheated bird will begin breathing with its mouth open and may lift its wings slightly to let cooler air up under its feathers. A juvenile bird might droop its neck over the edge of a nest.

Hyperthermia must be treated quickly. Animals should be shaded and spritzed with cool water, especially under their limbs. They may be provided with cool surfaces, such as wet towels, or towels packed with ice bags. Good ventilation and airflow will help animals expel heat. A very small mammal can be placed inside a plastic leak-proof bag that is then held in a tub of cold water. In severe cases, an animal can be submerged in cool water.

The stress of handling can exacerbate hyperthermia in many adult birds. These animals should be placed in a cooler environment and left alone to cool without handling. Any bird that is open mouth breathing should not be handled at all unless unavoidable. If you are unsure why a bird is open mouth breathing, stop handling it immediately and give it an undisturbed chance to calm itself.

Basic wound care

Occasionally a rescuer will be presented with a wounded animal that will benefit from having its injuries treated prior to arriving at a definitive care facility. As with any additional handling, the rescuers must weigh the benefits of extra handling over the pain and distress it will cause the animal. If their efforts increase the animal's chances for survival, then it is reasonable. This should ultimately be the decision of the veterinary overseeing the rescue efforts or receiving the animal.

When assessing an animal's wounds, try to establish the cause. Was the animal attacked by or observed fighting with another animal? Was it entangled in something such as fishing line that someone already removed? Was it shot? If the wound is bleeding, it can be controlled with pressure applied directly to the site when possible. Try to ascertain the wound's size and depth. Tourniquets are not recommended. Any restrictive wraps to control bleeding must be loosened within an hour. Neglecting to do this may result in the animal's death due to tissue damage.

There are three basic stages of wound management: cleaning, closing, and covering. Under ideal situations, the wound is protected by wet gauze or a water-based lubricant while the area around it is cleaned using a solution of povidone-iodine (Betadine), diluted to the color of a cup of tea, or chlorhexidine solution, diluted to the color of blue sky, or any wound cleaning preparation. After the area has been cleaned, the wound itself can be gently irrigated with a stream of sterile saline or diluted wound cleaning solutions, at room temperature or slightly warmer. The wound cleaning solution can be used to rinse the area using a 35 ml syringe with an 18 gauge needle to create a thin but forceful stream. Care must be taken not to drive dirt or debris further into the wound with too much pressure. Birds with deep or penetrating wounds should not have wounds flushed with fluid as there may be punctures into the airsacs and flushing the wound may drown the bird. This also applies to cleansing of fractures of the humerus or femur, as these bones connect to the respiratory system. Once at the definitive care center the wound should be debrided – cleaned of dead tissue and foreign matter.

If a wound requires extensive cleaning or suturing, it should be seen by a veterinarian without delay – ideally within six hours after injury. Until then, keep the wound moist and protected with bandaging. Semi-occlusive dressings such as hydrogel can be applied before the wound is bandaged. For most species, topical medication should be water soluble, such as silver sulfadiazine cream. Sugar or honey dressings are very useful for heavily contaminated wounds but are not appropriate for exposed bone or wounds that have a chance of being surgically closed when seen by the veterinarian, as the hyperosmotic nature of the dressing dries out the tissue at the edges of the wound.

Bandaging is usually composed of three layers, but the ideal bandaging is dependent on the species and the nature of the injury. Protecting the wound from further contamination is the ultimate goal of field bandaging.

The *contact (primary) layer* should be sterile and non-adherent, and provide moisture to allow the wound to drain yet stay moist. There may be situations, however, where a veterinarian will prescribe wet-to-dry dressing, in which case this layer will be wet and meant to adhere.

The *intermediate (secondary) layer* is applied on top of the contact layer to help absorb fluids, immobilize injury, and protect the area with thick cushioning material, such as cast padding.

The *outer (tertiary) layer* protects the other layers and holds them in place. This is usually made from a flexible, stretchy, self-adherent wrap.

Stabilizing fractures

Fractures require immediate attention and veterinary care. The wildlife medic can stabilize these injuries in the field, helping preserve the soft tissue and reducing further injury and discomfort during transport to a definitive care facility. Since wild animals must be fully functional to be released, fractures are considered more life threatening for these animals than for domestic pets.

Compound fractures, where a broken bone is exposed, must be kept moist to keep the bone viable. Exposed bone can be covered with gauze padding saturated with sterile saline solution. The pack can be secured in place with bandaging. The entire site should then be stabilized with the appropriate splint or wrap. Be careful not to allow the wrap to act as a tourniquet.

As a rule, to stabilize a fracture, the nearest joints above and below the fracture are immobilized. Leg fractures in mammals can be stabilized using a Robert Jones bandage. In birds, simply folding the limb into a normal position against the body and taping it in place using a plumage-friendly tape such as 3M Micropore Paper tape can be more effective as temporary immobilization than struggling to apply a Robert Jones bandage. If a leg is taped to the body, the bird will need a nest-shaped soft support during transport.

Robert Jones bandage

The Robert Jones is a bulky dressing used to immobilize fractures of the extremities. It is the bulk of the padding and how it is fairly tightly wrapped that immobilizes the break. The Robert Jones is often the first choice of bandage for mammals prior to transport.

To begin the application of a Robert Jones the leg should be resting in a normal position, slightly flexed. Two strips of tape are placed on the limb, anteriorly and posteriorly, extending several inches beyond the foot. As needed, gauze can be wrapped around the tape to secure it. Next, generous layers of cast padding or rolled cotton are wrapped firmly around the limb, distally to the most proximal point on the leg. The wrap, when finished, should appear about three times thicker than the unwrapped leg. The toes should be left unbandaged.

After the cotton padding has been applied, the tape "stirrups" can be folded upward. Conforming gauze is then wrapped fairly tightly around the padding, distal to proximal, as always. If done correctly, the bandage should be firm and even. When necessary, additional support can be added.

Finally, the bandage is covered with a tertiary layer of self-adhesive material. This, too, should be applied firmly so that when tapped with a finger it sounds like a ripe watermelon. The tape stirrups are used to hold the bandage in place.

Figure-eight wrap

A figure-eight bandage is typically used to stabilize fractures of a bird's radius, ulna, or metacarpals, or when those segments of wings are swollen in a manner that makes the rescuer suspect a fracture. An added body wrap holds the wing in place if it is drooping after application of the figure-eight.

Use self-adherent wrap for birds larger than 250 g and plumage-friendly tape such as 3M Micropore Paper tape for smaller birds that are likely to wiggle out of self-adherent wraps very fast.

To start, identify the elbow and wrist (carpus) of the bird. Sometimes fractured wings may be spun at the fracture site; make sure the wing is oriented correctly before wrapping it. Cut the amount of self-adhesive bandaging needed from the roll. It can then be sliced to the appropriate width and loosely rolled back up before being applied. An ideal width for most species is no wider than one half the elbow to wrist length.

Fold the wing into a normal position against the body. The wrap begins with placement of the bandaging around the folded carpal joint in the direction indicated in Fig. 71 (1). The bandaging is then unrolled over and then under the metacarpals (2), then under and over the elbow (3), making sure to include the tertiary (scapular) covert feathers and get the wrap well onto the humerus to ensure the elbow is included. The wrap should sit mid-humerus, above the elbow. Complete the "eight" by returning the wrap around the carpus (4) and traverse the entire path again.

Fig. 71 The figure-eight wing wrap.

The purpose of this wrap is to secure the wing in its normal position. If it has been applied too tightly or with the wrap material extending past the fleshy end of the wing tip, the secondary and primary feathers will appear criss-crossed. If the elbow is not well secured, the wrap may do more harm than good.

Bird body wrap

A fractured humerus or injury to the scapula or clavicle may present as a drooping wing. The wing can be stabilized with a body wrap. As previously mentioned, use self-adherent wrap for birds larger than 250 g and plumage-friendly tape such as 3M Micropore Paper tape for smaller birds that are likely to wiggle out of self-adherent wraps very quickly.

Apply a figure-eight wrap to the affected wing. The bird's injured wing should then be secured in place by wrapping the bandaging material around the body, behind the unaffected wing and in front of the legs, allowing those limbs free movement.

This wrap should be snug but not so tight that it interferes with the bird's breathing. To check to see if it is too tight, the medic should be able to fit two fingers underneath the bandage between the body and the wrapped wing. A well done wrap will allow the wing to be held symmetrically to the uninjured wing.

The ball bandage

The ball bandage is used to stabilize toe fractures in birds or to aid the healing of other serious foot injuries. Gauze padding or cotton is used to form a rounded pad on which the bird's toes are conformed, placed in a normal grasping position. The foot is then wrapped using conforming gauze, starting with a wrap or two around the ankle to secure it in place. The idea is to wrap the foot as evenly as possible, up and over and around. If done correctly it will take on the shape of a ball, with sufficient padding to allow the toenails to be oriented correctly, not twisted to the side. Self-adhesive bandaging can be applied as the tertiary layer.

13 Transporting wildlife

For a wild animal, transport can be an incredibly stressful and physically challenging experience. The vehicle's acceleration and braking, and curves in the road, can force the animal to shift its weight and steady itself. This might be excruciating for an animal with a fractured limb, even if it has been stabilized. Loud and unfamiliar sounds add to an animal's level of stress. Even the quality of road has been shown to impact animals during transport. Rescuers must prepare for the animal's welfare by choosing suitable caging and the most appropriate mode of travel.

Transport can come in all sorts of shapes and sizes, from compact automobiles to cargo planes. The key factors that will influence the animal's physical wellbeing during travel will be the ambient temperature, airflow, air exchange, and the substrate of its carrier.

As a rule, wild animals should not be transported in an open bed of a truck. Inside a transport vessel, the ideal range of temperature for most animals will be between 65 and 70° F (18–20° C). Marine mammals may need to be kept cooler, closer to 50° F (10° C). If animals are young, emaciated, or hypothermic, the temperature should be higher. Highly stressed and active animals, especially those with thick pelage or dense feathering, may need to be kept cooler.

Regarding airflow and air exchange, the transport vessel should be equipped with an air conditioner that can keep a constant exchange of fresh, temperature controlled air flowing throughout the vehicle. At the very least, there should be some method of delivering fresh air. In an enclosed cargo van, for example, PVC drainage pipes can be used to deliver air from the outside. The drainage pipe should be fitted with an elbow and extension so that it hooks around the back of the vehicle, facing forward, so that fresh air is forced inside while moving. It is important to make sure air can also escape. If the van is equipped with a sliding rear door, a block of wood can be used to create a gap. This air exchange also helps to keep engine exhaust from entering the rear cargo area.

When transporting more than one animal, animal carriers should be spaced so that there are at least two sides exposed to airflow, with a minimum of 2 inches (5 cm) between boxes for adequate ventilation. Transport boxes should never be stacked right next to one another without space in between.

Wildlife Search and Rescue: A Guide for First Responders, First Edition. Rebecca Dmytryk.
© 2012 John Wiley & Sons, Ltd. Published 2012 by John Wiley & Sons, Ltd.

Transporters should listen out for sign of distress or discomfort, which include shuffling, thumping, and constant vocalizing. Animals should be checked every 40 minutes or so during long journeys. Birds that are overheating will exhibit open-mouth breathing and may elevate their wings slightly. Signs of hyperthermia in mammals include rapid breathing and panting.

Some animals will be more comfortable traveling with conspecifics. Pelicans, for example, travel well in groups, as can elephant seal pups. Tarps and sheets can be used to line the rear compartments of vehicles, creating a spacious environment. A PVC frame and prefabricated material can be built ahead of time, making conversions quick and simple.

While some gregarious species of birds will do well when housed in close proximity, others, such as western grebes, will not. When in a stressful environment, normally tolerant species can become violently aggressive toward conspecifics. The decision to house conspecifics together should be left to someone with experience. In general, the following species have been successfully transported in groups: pelicans, common murres, guillemots, auklets, horned grebes, eared grebes, most duck species, mergansers, geese, swans, penguins, terns, skimmers, sandpipers, avocets, stilts, and American coots.

When transporting birds on perches, transporters should try to position the carrier or perch so that when the vehicle accelerates or slows the bird only needs to shift its weight. In other words, face perching birds perpendicular to the vehicle's momentum.

14 Field euthanasia

Any animal suffering from severe injury or disease constitutes a veterinary emergency. Wildlife rescuers have an ethical and legal obligation to provide for these casualties appropriately and promptly. Occasionally, this may include euthanasia, or mercy killing. Euthanasia is the act of relieving pain and suffering by ending the life of a hopelessly ill or injured individual in a relatively painless manner. The word is a combination of Greek words, "eu," which means well, or good, and "thanatos," meaning death.

For wildlife rescuers and rehabilitators, euthanasia is a means of relieving an animal from pain and terror. Euthanasia should be considered for animals that are in severe, irremediable pain or distress and for animals with injuries that prevent them from surviving in the wild. Euthanasia should also be considered appropriate for an animal that might be a candidate for life in captivity, but where its quality of life would border on cruelty.

More often, the fate of a wounded animal will rest with the wildlife veterinarian or licensed wildlife rehabilitator, not search and rescuer personnel. However, there may be times when the decision will be theirs. This chapter is offered for situations where wildlife search and rescue personnel are faced with critically wounded animals – animals that are still alive, suffering, and stand no chance of surviving, and where transport would inflict greater pain and suffering.

The decision to take an animal's life should be based on policy and sound judgment. However, coming to terms with killing an animal is a personal journey. People's perception of euthanasia, it being right or wrong, appropriate or inappropriate, is deeply rooted in their belief systems and as such they should not be criticized for what they believe in.

Methods used to end an animal's life must be based on humane guidelines. The aim is to bring immediate relief from pain and suffering through rapid loss of consciousness and death with minimum pain, discomfort, and distress. Methods will vary depending on the type of animal, the means of euthanasia available, and the rescuer's level of abilities and training. Electrocution, exsanguination, drowning, air embolisms, anoxia and asphyxia, and exposure to extreme temperatures are, by themselves, considered inhumane and are not recommended for conscious animals.

Wildlife Search and Rescue: A Guide for First Responders, First Edition. Rebecca Dmytryk.
© 2012 John Wiley & Sons, Ltd. Published 2012 by John Wiley & Sons, Ltd.

The following methods of field euthanasia are offered as alternatives to end an animal's suffering when veterinary assistance with controlled drugs is not available or when transport would cause the animal an unacceptable amount of pain and distress. Field euthanasia should not be performed in public or with an audience.

Destruction of the brain through cranial trauma, when done correctly, is the simplest, quickest, most humane method of primitive field euthanasia. The objective is to cause immediate loss of consciousness followed by death. In small to medium-sized animals, this can usually be achieved with a swift blow to the back of the skull using a blunt instrument. The action must be carried out with as much force as possible. The death of a stunned animal must be assured by the use of an additional technique: for example, exsanguination, asphyxia, decapitation, or repeated blows to the back of the head to completely destroy the brain.

In preparation, the head of the animal can be covered with a lightweight towel or tissue. The head should be resting on a hard surface, nothing that will give in, and it should be positioned so the handler can make an accurate and powerful strike on the first attempt. Handlers should wear appropriate safety gear and be prepared to dispose of the carcass in an appropriate manner, consistent with local, state, and federal regulations.

A manual blow to the head may not be effective on larger animals with thick skulls; it may require either a bullet or a captive bolt. If it is legal for a firearm to be discharged, this can be a quick and effective means of ending an animal's suffering. Where discharge of a firearm is illegal, a cartridge-fired captive bolt pistol can be used to stun an animal. Once the animal has been correctly stunned, death must be ensured by, for example, exsanguination. Carers must make sure the caliber of bolt and the strength of cartridge are appropriate for the animal they are assisting.

Decapitation is not typically a method of euthanasia by itself. However, in very small mammals, if done swiftly and effectively, loss of consciousness is rapid, even though electrical activity may continue in the brain for seconds after. Decapitation is not suitable for birds as visual responses can continue for up to 30 seconds after decapitation. Decapitation is also not suitable for reptiles or amphibians unless the animal has been rendered unconscious beforehand.

Carbon dioxide is suitable for the euthanasia of small mammals and small birds. Compressed CO_2 gas cylinders are used to deliver a regulated flow of gas into a deep container, such as a plastic bin. The container must be able to be filled enough that the animal cannot raise its head above the gas.

Conscious animals can suffer severe dyspnoea and distress when placed into a chamber containing a high concentration of the gas. Therefore the gas ratio should be increased over time to the appropriate amount that will induce loss of consciousness and death with little to no distress. To ensure death, they can be exposed to 100% CO_2 for an additional 5 minutes. When there is doubt that death has occurred, a physical method should be utilized.

Gas agents normally used to anesthetize animals are often utilized for euthanasia. Halothane is one of the preferred gaseous anesthetics, followed by enflurane and isoflurane. Isoflurane, however, has a pungent odor and thus may not be suitable for use on a species with a large breath-holding ability. Administration of inhalant anesthetics, by or under the direction of a veterinarian, may be performed by placing an animal in an airtight container with material that has been soaked with an appropriate amount of the anesthetic. Animals must only be exposed to the vapors. The chamber must also contain a sufficient amount of air or oxygen to prevent the animal from experiencing distress as it loses consciousness. Specially made chambers provide a means of slowly closing off the source of fresh air as the animal succumbs to the anesthetic.

Inhalant anesthetics pose serious heath risks to humans. These risks must be considered when wildlife rescue organization prepare their personnel for field euthanasia, especially if they will be working in remote areas where accidental exposure to these potent agents could be deadly.

15 Life, liberty, and euthanasia

Wildlife rehabilitation is the nursing of wild animals so that they may be returned to freedom. However, there are times when patients are so badly injured that their survival in the wild is questionable, if not highly unlikely. For the wildlife rehabilitator, the choices for what to do with an animal include humane euthanasia, placement into a captive setting, and, on very rare occasions, release.

There are many cases of animals living wild and free, thriving despite a physical impairment. There are documented cases of deer, fox, and raccoon surviving with only three legs. There are numerous accounts of one-legged gulls, waders, hawks, passerines, and even shorebirds, not just surviving, but raising young. So, every once in awhile, rehabilitators will have a slightly impaired animal that they believe stands a chance to see out its life wild and free.

Wild animals are meant to live free. It is their nature. For a wild animal, quality of life means it is at liberty – able to make all its own choices. Captivity is no place for a wild animal. No matter how much room it is given, no matter how much enrichment it has, no matter how much its enclosure resembles its wild home, there will always be one thing missing – the thing it wants most – its freedom.

Ever so often, wildlife search and rescue personnel will also encounter wild animals that are impaired, but surviving. There will be a question as to what is in the animal's best interest. Should it be put through the terrifying ordeal of capture only to be euthanized? Should it be retrieved from its wild state only to be placed into captivity?

Anecdote: The story of Peg

Occasionally, rescuers will encounter permanently maimed wild animals that are eking out an existence – adapting to their impairment. This is the story of that lucky plover and her rescuer who thought outside the box, past the rules, and beyond euthanasia.

One day a researcher came across an injured female Western plover – she had a nearly severed foot. He captured her and called around for help from rehabilitation facilities. After describing the injuries they all said she would be euthanized because that's what the book says to do. It's the law. The researcher finally found a wildlife facility, miles and miles and many counties away, that said they would at least try to reconnect the dangling appendage.

Days later, despite all of their great efforts, the foot had to be amputated. She was not killed but kept alive at the request of the man who had found her. The researcher figured he owed her at least that, to die free.

Weeks went by and the little plover had gained in strength and was ready to be set free. The researcher drove the little bird back to her home territory and released her exactly where he'd found her. She was quick to take flight out of the box and landed some distance down the beach.

The researcher came back every morning to look for her. He saw her for a few days and then she vanished.

Months later, as the researcher was surveying the mouth of a riverbed, he noticed peculiar tracks in the mud bank – one plover footprint and then a hole, one plover footprint and then a hole. It was Peg! He followed the tracks as they led inland, far from the shore, over gravel bars of rounded stones and cobble, and then he spotted her, being mounted by her longtime male companion.

After that encounter the researcher met up with Peg many times. She lived at least three years and successfully hatched and raised multiple broods.

16 Rescuing baby birds and land mammals

In spring and summer months, wildlife rescuers will experience an increase in calls, most of which will involve young animals. First and foremost, the rescuers must judge whether or not intervention is necessary. They will rely on their familiarity with the species – its natural history – in assessing the animal's condition. In general, baby birds that are fluffed up, are not vocalizing or gaping, and have a "sad" look need immediate help. Infant mammals that are cold to the touch and curled in a fetal position – not moving much, even when touched – also require immediate aid. These babies need to be warmed, and quickly.

An incubator would be the best choice for providing warmth. Commercial or homemade incubators will provide babies with the heat and humidity they need. The temperature requirements will vary depending on the species and age, but in general, the ambient temperature will need to be between 85 and 95° F (29.4–35° C). In the absence of an incubator, rescuers can devise a makeshift heat source as discussed in Chapter 12 in the section on treating hypothermia.

As a young animal warms, it will start to move, and possibly to vocalize. If they have not done so yet, rescuers will want to do a brief, cursory exam to make sure there are no hidden injuries, such as puncture wounds, or excessive ectoparasites, such as fleas and ticks, or the presence of fly eggs. They will also check for signs of illness. Drainage from the eyes or nose can indicate a problem. Rescuers will also assess the animal's level of dehydration. While the animal's life-threatening needs are tended to, the rescue team should consider contacting the wildlife rehabilitation facility that will be receiving the animal and take instruction from them on how to proceed as far as any further care is concerned.

When a young animal shows no sign of injury or illness and is in good condition, keeping it wild, with its wild family, should be the ultimate goal.

Wildlife Search and Rescue: A Guide for First Responders, First Edition. Rebecca Dmytryk.
© 2012 John Wiley & Sons, Ltd. Published 2012 by John Wiley & Sons, Ltd.

17 Reuniting, re-nesting, and wild-fostering

A young wild animal stands a greater chance of surviving as an adult and leading a normal life if raised by a wild parent. From wild parents, young learn where to forage and hunt, what to eat, what to be afraid of, and where to find shelter. Growing up wild they also acquire valuable and necessary social skills. Wildlife rehabilitators are able to rear most species from infancy to adulthood but their nurturing will always fall short of what a wild parent would provide. Therefore, it is only right to see that healthy wild babies remain in the wild whenever possible. This must be seen as part of the duty of a wildlife rescuer.

Those who work in the field of wildlife rescue and rehabilitation are expected to adhere to a set of standards and a code of ethics in providing the best achievable care to their patients. If, then, being raised in a wild family unit is what is best for a healthy wild animal, then it is a wildlife rehabilitator's obligation – it is their responsibility – to make every effort to return babies when possible.

Adopting a policy to return healthy babies to the wild also benefits the rehabilitation program. During spring and summer months, wildlife rehabilitation centers can become overcrowded with patients, many of which are healthy babies. By dedicating a small amount of resources on a program that returns them to the wild, admissions can be dramatically reduced. One such program reported a 29% reduction in healthy baby bird admissions during their first year.

To begin providing this valuable service, a rescue organization will want to review its admission policies. Wildlife centers that receive animals from outside sources, such as veterinary clinics or shelters, should start to require complete finder information on every animal brought in, making it a condition for receiving animals into care. This information is so important, so crucial to the process of reuniting healthy babies, that tough measures like this are warranted. Conditional admission policies have proven extremely successful in cutting down the number of animals that come in with incomplete paperwork. Occasionally, animals are "dumped" at a facility, leaving no clue as to where they were found or by whom. If a wildlife center receives too many "dumped" animals from a single source it might want to stop taking the animals from them until they tighten up their own admission policies. It is that important.

Wildlife Search and Rescue: A Guide for First Responders, First Edition. Rebecca Dmytryk.
© 2012 John Wiley & Sons, Ltd. Published 2012 by John Wiley & Sons, Ltd.

The next step to forming a reuniting program involves structure and preparation. The organization will want to write a kind of business plan, detailing the program's objectives and how it will operate. Will it have a limited service area or limit the species it works with?

In this planning stage, the organization may want to contact other groups that offer this service and see if they would be willing to share information. It will also want to start collecting information on reuniting and wild fostering, building a reference library for personnel.

Ideally, the organization will have a dedicated team of experts to receive calls about uninjured baby wild animals. The primary goal will be to reunite healthy babies with their families or to place them into an acceptable wild foster family situation. Their job starts with the finder.

It is important that rescuers try to contact the finder as soon as possible while details are still fresh. Rescuers will want to collect a detailed history, including where, exactly, the animal was found, when, and under what circumstances. They will ask whether parents were observed. They should also find out if the animal was given any food or water while in their care. The rescuer will use all this information to make a preliminary assessment of the animal's condition and whether or not it makes a good candidate for reuniting. The following preconditions must be met for an animal to qualify.

Rescuers must be able to correctly identify the species and its age or stage of development. The animal must appear healthy with no apparent signs of injury or illness, and stable enough to withstand the process. Birds that will be in or near water must be waterproof and their feathers must be free from contamination of any kind.

Another important provision is that the person attempting to return a baby to the wild (or overseeing the process) must be familiar with the natural history of the species involved. They should also be experienced with the process or be under the guidance of someone who is, even if their only means is via the phone or the Internet.

There is an old wives' tale that warns that handling a baby bird will cause it to be rejected by its parents. This is a myth. Other than vultures, birds do not have a highly developed sense of smell. They are more apt to key on visual changes – alterations to their nest, for example. However, the parental bond is so strong it seems to override concern over most alterations, even quite dramatic changes. Therefore, there should be no fear in touching a baby bird from that standpoint. However, for the baby's sake, the handler's hands must be clean, dry, and free of oil or dirt.

There are two main classifications of baby birds, based on their degree of development when hatched – precocial and altricial (Fig. 72). Precocial chicks are born with their eyes open and covered in down. They can walk soon after hatching.

Fig. 72 Precocial and altricial chicks.

Most altricial chicks are born naked with their eyes closed. Unable to stand or walk, they are heavily dependent on the nest and their parents. As they grow, hatchlings move through different stages of development. Once nestlings start to explore their treetop world, they can be called branchers. When their feathers are developed enough for short flights they enter the fledgling stage. This is when they fledge from the nest, taking their first flight – which is often down.

Returning altricial chicks to the wild

Initially, fledglings are not strong flyers, often spending a great deal of time on the ground, hopping about and perching on low branches. Some birds will be grounded for days, even weeks, before they are strong enough to fly. This is a precarious time for young birds, but it may be the most critical stage in their development.

While fledglings explore their world their parents continue to look after them, feeding and protecting them, even on the ground. Parental dependency varies. Some species, such as great horned owls, can be dependent on their parents for three months after they fledge.

Having a wild parent as a guide during this phase of their life is key to a bird's survival. During this stage, young birds become familiar with the landscape. They develop their hunting and foraging skills and learn what to eat, what to be afraid of, where to hide, and where to roost. If they are to disperse or migrate, those raised by wild parents will be much more prepared than birds raised in captivity. It is therefore imperative that healthy fledglings be returned to their nest territory whenever possible, even if they have been in treatment for injuries or illness.

When rescuers are presented with a healthy brancher or fledgling, they should first see to its immediate needs before returning it to the nest site. They will want to perform a thorough examination, looking for signs of injury or illness. The abdomen should be a healthy pink color and round, like a potbelly, not overly wrinkled or sunken in. Their eyes should be bright and round, not squinty or sunken looking. Keeping control of the bird's head if necessary, rescuers can softly blow through the feathers to look for hidden signs of injury such as puncture wounds, bleeding, bruising, or torn skin. Their wings should be equal in appearance, held up fairly close to the body, not dragging. When they are gently extended, the baby should be able to pull them back into proper position easily. Although it might wobble, a fledgling should be able to hop, walk, and perch. When they are gently extended, the baby should be able to pull its legs back into normal position. Rescuers will consider its temperature. It should feel warm to the touch. If the baby is puffed-up looking, it is probably cold. Its behavior must also be evaluated. Does it appear uninterested in its surroundings, preferring to tuck its head between its feathers? Does it look weak or tired, closing its eyes often? In considering whether a fledgling is healthy enough to withstand the process of being reunited, rescuers will also want to consider the circumstances under which it was found.

If the baby is injured or appears weak, ill, cold or de' ydrated, it should be seen by a wildlife veterinarian or licensed rehabilitator. It n ·y just need a day or two to recuperate before being reunited.

For fledglings or branchers that are healthy enough to be returned, the rescuer, guided by the finder's information, will want to return to where the bird was found to look and listen for signs of a parent. If it is a diurnal species, they will want to go during the day, and at dusk if it is a nocturnal species.

If there is no sign of adults in the area, rescuers may want to try to draw them in with calls. Taking the young bird and letting it hear and see its familiar surroundings may encourage it to vocalize. In some situations, recorded calls can be used. Sometimes the parents are just waiting for the humans to leave before they make an appearance. Rescuers may want to chance placing the young bird on the ground or a low branch without a confirmed sighting and watch from afar.

Fledglings and branchers are ready to be out of the nest and exploring. They can be restless, and don't stay put in one place for long. Fledglings and branchers should not be re-nested. Doing so risks disturbing any remaining nestlings and subjects the young birds to another fall.

Fledgings can be placed in a safe spot on the ground, under a bush, or on a low branch. Branchers should be placed on the highest branch that can be reached safely, keeping in mind that the brancher may not readily take to the perch. It may be best to place it next to the trunk of the tree where it will be less likely to fall if frightened. Ideally, the tree will have plenty of branches and good cover.

After setting the fledgling or brancher in an appropriate spot, the rescuer will want to retreat to a concealed location to watch for the reunion. Disturbances should be minimized. If people have gathered, they should be asked to watch quietly from a distance. Even in the most ideal situations, it will take time. The rescuers should give it at least an hour.

Reuniting nestlings can be a bit more complicated. It requires locating the nest and may even call for creating an artificial one. Nestlings also tend to be more prone to injury.

When presented with a nestling, rescuers will want to perform a thorough examination of its body, looking for signs of injury or illness. Nestlings are typically heavy-bodied. With undeveloped wings and legs, when they fall, they fall hard. Rescuers will look for evidence of blood, torn or punctured skin, bruising, awkward recumbency, abnormal or labored breathing. They will check the bird's temperature – it should feel warm to the touch. The nestling should have a big pink potbelly. The skin on the abdomen should appear tight, not overly wrinkled. If the eyes are open, they should be bright and round, not squinty and sunken in.

Rescuers will also consider its behavior. Is it responsive and vocal, or does it appear weak, keeping its eyes closed and head down? The circumstances under which it was collected will also be considered.

If the baby appears injured, dehydrated, cold, or weak, or has been caught by a dog or cat, it should be seen by a wildlife veterinarian or licensed rehabilitator.

Once the rescuer determines that the baby is healthy enough to be re-nested, they will direct their attention to the nest, if they can find it. Nestlings do not land far from the nest – they are usually found on the ground directly below the nest site. To find the nest, it will help to know the species. Will the nest be in a tree, a cavity, or a burrow? The baby bird may hold the clue. The young of cavity nesters will tend to orient themselves facing one direction and gape horizontally, whereas cup nesters will gape vertically and be arranged more randomly. If the nest is not readily apparent the rescuer will need to be quietly observant, waiting for the parent to return for the next feeding.

Once the nest is located, the rescuers must consider the cause of the displacement. Is the nest weak or damaged? Was the accident weather related?

If the nest is accessible, appears to be well anchored and in good shape, the nestling can be returned. Re-nesting should occur during the species' normal active time of day. In other words, if it is an owl, rescuers will want to wait and place the owlet into the nest just before dark so it doesn't go too long without food.

When re-nesting, if there are other nestlings still in the nest, and they are old enough to be fearful, caution must be taken to prevent them from falling out when the nestling is introduced. The rescuer should move slowly and quietly, keeping concealed from the nestlings as much as possible, reaching just over the edge of the nest to replace the baby. Once the baby feels securely within the nest, the rescuers can slowly remove their hand.

If an entire nestful of older frightened nestlings are being replaced, they can be introduced to their nest while under a dark colored towel. They should stay covered until the rescuer is ready to leave. Once the rescuer is as far away or out of sight as necessary, they will want to gently pull the towel away.

When a nest is located high in a tree, higher than an extension ladder will reach, there are some options. There might be a local arborist willing to help climb the tree. Another option would be to find someone with a "cherry picker," as used by fire departments, utility companies, tree trimmers, and some sign repair shops.

If the nest is in a precarious location, out of safe reach, rescuers may need to consider trying a reunion after the babies fledge. If the nest site has been completely destroyed, rescuers will need to consider building a new nest site in close proximity.

Nests come in a variety of shapes and sizes. Some are simple scrapes in the sand, with little nest material. Others can be splendid, elaborately woven pouches. When retrofitting a nest, or replacing it altogether with a manmade version, rescuers will want to use as much of the original nest as possible, reproducing its shape, color, and design. As a rule, nests should be situated as close to the original nest site as possible. Once it is securely in place, rescuers should set the baby bird in the new nest and leave quickly. They will need to move far away and watch for the parents' return from a concealed location.

There are a couple of rules for makeshift nests. They must not be made of things that can harm the baby birds. Fraying fabric, thread, or fabric with tiny loops should be avoided as toes and legs can get twisted up or snagged. Ideally, replacement nesting material should be natural and from the surrounding area, like dry grasses, but thinly shredded unbleached paper towels can work. So long as it doesn't retain moisture.

The depth of a cup nest is also important. The ledge must not be too high – it must resemble the original nest. Baby birds will lift their bottoms up to eliminate on or over the rim of the nest. If it is too high, they will soil themselves and the whole interior of the nest. This can lead to feather rot and disease.

Small to medium-sized cup nests are fairly simple to retrofit. There are many common household items that can be suitable to use as a base – small fruit baskets, gift baskets, or a shallow plastic container, such as a margarine tub, or a funnel. When using a plastic tub, holes must be punched out of the bottom so that it will drain. A couple of holes can also be punched on the rim to help secure it. Once the nest is secured in place, rescuers need to double check that it drains sufficiently and not hold water.

Tiny cup-like nests, such as a hummingbird-sized nest, can be formed out of a professional microphone cover. The center of the foam unit can be pushed in from the top until it is almost completely inverted. This should form a nice soft, elastic cup. The edges should be secured so that it retains its shape. Another method for creating a small cup involves cutting the top off of a small plastic drink bottle. Holes should be drilled in the cap for drainage.

Larger nests, such as those for a tree-nesting owl or hawk, can be made out of a laundry basket, which will serve as a durable frame. Holes must be drilled in the bottom so that it will drain. Branches should be secured to the rim of the basket to eliminate any slippery surface. Twigs and branches can be woven through the laundry basket's weave, creating the nest itself. It should be built up to just under two-thirds of the way inside the basket. Dry pine "straw" or dried leaves from the original tree can be stuffed into the gaps between the twigs and branches.

When returning cavity-nesters, if the nest cannot be found, a nest box may be offered as long as the rescuer feels confident they have all of the babies in hand. Otherwise, the parents may ignore the new nest while continuing to care for their remaining young in the original nest.

The size of the box and size and placement of the entrance hole will vary for different species. In general, the box should be large enough to house a clutch of young comfortably, but not be so large that the babies become separated and chilled.

That being said, barn owls (*Tyto alba*) do better with larger sized homes. Additional room for the developing chicks will help reduce the chance of cannibalism and allow for a larger brood. Also, it is very important that the entrance hole be placed at least 11 inches (28 cm) from the nest floor. This helps ensure adequate development before owlets access the hole. If it is too low, a heavy-bodied downy owlet will risk falling to its death. Where the hole is above 12 inches (30 cm), cleats can be attached to the inside wall underneath the hole, but no closer to the entrance edge than 8 inches (20.3 cm). See Appendix 5 for instructions on building and installing a barn owl box. Being ledge-nesters, barn owls sometimes choose precarious ledges on which to raise their young. Without a raised border, the owlets are at risk of falling.

Anecdote: Return of baby barn owls

It was a cold and rainy Friday in Monterey, California. WildRescue received a call about a nest of owls that had been found in a large metal sign at an auto body shop in downtown Santa Cruz. The sign was being dismantled for safety reasons.

Duane Titus and Klaus Kloeppel arrived on scene within minutes, only to find that a nearby wildlife rehabilitator had picked up the babies, intending to foster them. In spite of the rehabilitator's resistance to reuniting the owlets with their parents, Duane and Klaus constructed an owl box on top of the pole that once held the sign.

That night, the parent owls were documented near the box. Both male and female were seen circling with large rodents in their talons. One landed on a nearby street lamp, a gopher dangling from its beak. Sadly, the rehabilitator had refused to return the babies that evening.

Finally, after two days of relentless persuading the rehabilitator agreed. The babies were placed in the owl box before sunset. The parents circled, clicking. The owlets began hissing and screeching, drawing the parents closer. At last, a couple of hours later, one of the parents made a food drop inside the box.

The next day the owlets were checked on to see that they were in good shape. They were each offered a couple of mice in case they had not received a meal the night before. This was done before noon so the chicks would be hungry and vocalizing by nightfall. The babies were monitored the following day and then, since we were convinced the parents were caring for them, they were left on their own.

Weeks went by and we received a call that the owls had fledged. All five babies could be spotted in the evenings, calling to their parents and practicing flight. With our presence we were able to involve the many locals who were curious about their wild neighbors. We explained how important this stage is to birds – the fledgling stage. This is where they learn where to hunt, what to fear, where to roost – things that no human can teach them. We also explained the value of barn owls. They are increasingly popular for rodent control as they are known to consume twice as much weight in prey as other owls. For instance, a maturing barn owl baby may consume the equivalent of a dozen mice in one night.

Returning precocial chicks to the wild

The most important action to take before reuniting or wild-fostering precocial young is to evaluate a bird's waterproofing. Improper handling and husbandry can soil a young bird's insulating down. If the bird is to be returned to the wild, its waterproofing, in most cases, must be excellent. The bird must be able to repel water and float – its feathers should not soak up water. If necessary, the bird may need a quick bath to get rid of soiling.

Some species, such as gulls and terns, go through a semi-precocial stage where they are physically capable of standing and walking but they stay near the nest to be fed by their parents. These species will lose much of their down before venturing into the water so, while it is not as important for these species to be 100% waterproof, rescuers should still consider their exposure to cold wind, rain, and direct sun. In the wild, a chick may become soiled with excrement but the downy feathering on its head, back, and wings keeps it sheltered from the elements.

Rescuers must also take into account the unique traits of the species they are dealing with. They must have knowledge of the different stages of parental care and chick development to successfully return young to the wild. Again, as an example, gulls and terns can nest in large colonies with hundreds, if not thousands, of other nests. As the chicks develop, they start to mingle together, forming what is called a crèche, or public nursery. In this type of situation, it is the distinct call of its chick that allows the parent to locate it among the others. For this reason, a healthy gull or tern may be reunited not by entering the colony but on the outskirts. Entering a breeding colony to return a displaced chick is unadvisable – the risk of disturbing the colony and displacing other chicks is far too great.

Young galliformes can be fairly easy to reunite as long as the chick is kept warm and healthy during its time away from its parents. Typically, the adults follow a foraging pattern, visiting the same sites daily. If the finder can lead the rescuer to the exact location where the baby was picked up, there will be a good chance that the parents will return. These species tend to be skittish. The baby should be placed in a small cage that can be tipped or opened from a distance. It should be situated near where the birds travel. If the weather is cool, the baby may need supplemental heat while it waits for its parents' return. The rescuers will need to stay quiet and out of sight. Once the parents are spotted, the chick should be set free. The chick's peeping and distress calls will alert the parents and help them locate their baby.

Most ducks, geese, and swan species can be reunited as long as the reunion takes place within a day or so of the displacement. If too much time goes by, some species, such as mallard ducks, may treat the baby as though it were an intruder. Other species though, such as Canada geese, will readily accept babies that are unrelated.

Some species, such as Canada geese, form groups, or crèches, of unrelated goslings. This is called brood amalgamation. They will often accept new, unrelated goslings as long as they are about the same size as their own. Preferably, introductions should be made before the goslings are one month old. As a rule, the goslings being introduced should never be smaller than the ones in the crèche.

To reunite or foster goslings into a crèche, rescuers must first locate a suitable group. Next, they will want to approach the group to within about 20 yards (18 m) if possible, without disturbing the wild family.

Because goslings key on movement, it is important that the movement they see is the crèche, not the rescuers. Ideally, rescuers will set the carrier of orphaned goslings on the ground, facing the crèche. Hiding behind the carrier, rescuers will then quietly open the door, setting the goslings free. It is more than likely that the goslings will head straight for the wild flock. The adults may react inquisitively at first, before signaling a greeting in goose-language.

Returning baby mammals to the wild

Young mammals can also be successfully reunited with their parents as long as they are warm and healthy. When presented with a baby mammal that has no apparent injuries, rescuers will need to examine it thoroughly.

To check for dehydration, rescuers can gently pinch the loose skin behind the animal's neck. If it remains "tented," the animal is dehydrated. The color of urine can also be an indication of dehydration. Wearing protective gloves, rescuers can gently stroke the genital area of a small mammal to stimulate elimination. If the color of the urine is brown, or gritty, the mammal needs immediate care. Rescuers also need to check the anus to see whether it is blocked, and wipe any dried feces from the area using a warm, moistened cloth or tissue. Every mammal that does not yet have its eyes open must be stimulated to urinate and defecate.

Warmth is extremely critical. If the baby is cold to the touch and sluggish, it requires warmth immediately. Just as with baby birds, baby mammals are unable to thermoregulate and can easily become chilled. Baby mammals must be kept warm throughout the reuniting process. Rescuers must be prepared to provide adequate warmth prior to and throughout the process in most cases.

While mild levels of dehydration and hypothermia can be treated in the field under the direction of a licensed rehabilitator, severe cases will require more extensive care. If after treatment the wild baby is healthy enough to be returned to its family, a reunion should be attempted.

More often than not, rescuers will be presented with babies that are perfectly healthy – victims of well meaning finders who are unaware that many animals will leave their young for hours at a time. Deer, for example, will leave their young hidden in tall grass or brush. A doe will not stray far; she will be close by, watching and listening. Sometimes fawns will hear people walking past and they will come out from hiding. Not realizing that the mother is nearby, people often assume the fawn is abandoned.

Fawns can be successfully reunited up to three days after being separated, though the sooner the better. What is important is that the fawn is healthy and has not been fed anything that might make it sick, such as cow's milk. As for being concerned about smells, fawns have been successfully reunited after being licked by dogs and coddled by children.

Reunions can be attempted in the day or night by returning the fawn to where it was initially found. Rescuers should leave the vicinity immediately. If the doe is there, she is watching and waiting until all humans are out of sight.

Cottontail rabbits nurse their babies for just a few minutes in the early morning and again in the evening. After nursing, a mother rabbit will leave her young in a type of nest – a small depression that she has made in grass, lined with her fur and covered with a light layer of leaves and grasses. Often these nests are right out in the open, easily disturbed by people who are mowing or raking their lawn.

When presented with a nest of baby rabbits, the rescuer will want to look for signs of illness, injury, or true abandonment by the mother. Baby rabbits that are warm and wriggly, with full bellies that are pinkish in color, not blue, should be replaced in their nest at once.

To check whether the mother has returned to feed them, rescuers can set two pieces of string on top of the nest in an X pattern. After about eight hours, if the string is not disturbed the mother may not be visiting. Rescuers should double check by looking at the babies' bellies – if they are still full and round and the babies look healthy and are not crying constantly, the mother has probably slipped in.

Young cottontail rabbits begin exploring outside of the nest by about two weeks of age when they are just 5–6 inches (12.5–15 cm) long. Cottontails that are about the size of a tennis ball are old enough to be on their own and can be returned to where they were found to rejoin their family.

Healthy baby tree squirrels can also be reunited, even if their leaf-nest, or drey, has been destroyed – often the mother has a second nest. Healthy babies can be placed in a shallow box, basket, or tub lined with dry grass or dried leaves from the immediate area. The container should be just deep enough that they cannot crawl out. If it is too deep the mother will be unable to retrieve them. If the baby squirrels still have their eyes closed, they might need a source of warmth, such as a bag of microwaved dry rice.

There are a couple of options for reuniting the babies with their mother. The container can be left at the base of the tree from which they fell or an adjacent tree. Another option is to secure the container in a tree where she might access them more easily. Once the container is set, rescuers should retreat to a hidden vantage point to watch for the mother's return. This may take two hours or longer. To hasten the process, an audio recording of a baby tree squirrel in distress can be played. If the first reunion attempt is unsuccessful, it can be attempted the next day as long as the infants are kept warm and fed (under the strict guidance of a licensed rehabilitator).

To reunite rats and mice with their mothers, the babies should be left in their nest material or something similar, such as unscented shredded tissue. The bundle can be placed in a very small, ventilated box tipped on its side. A heat source, such as a gel pack or heating pad, placed under or against a portion of the box should keep the babies from wanting to stray. If they are older, and very active, they may just need to be kept warm and hydrated until nightfall when they can be released where they were found.

Larger mammals can also be successfully reunited. Raccoon mothers have reclaimed their young even after three days. If the babies are too young to be roaming with their mother they can be placed into a plastic tub or nest-box. An insulated nest-box can be made out of two cardboard boxes, one smaller than the other. A heating pad can be placed under a portion of the inner box. Shredded paper can be used to insulate between the boxes. Non-fraying fabric, such as flannel, or shredded newspaper can be placed in the inner box. The nest-box should be situated in a quiet protected area, near a tree or fence – somewhere along the mother's route where she is likely to find it. A small sheet of plastic, such as a garbage bag, placed underneath the outer box will help keep the bottom dry. Using gloves, the warm cubs can be placed into the nest-box. The flaps of the cardboard box can be allowed to fall inward, as if shut. This will help hold in heat but not prevent the mother gaining access.

To assist the mother in finding her young, rescuers can gently rub the babies with a fragrance-free towel or tissue to collect their scent. These can be used to guide her to the container. In some cases, it might be kind to provide her with sustenance, such as a bowl of water or a few grapes set outside of the nest-box.

The process of reuniting raccoons can take time. Even when the mother finds her babies quickly, she may not retrieve all of them at once. If a few of the babies are still in the nest-box the following morning, they must be kept warm and hydrated. If the young require bottle-feeding they can be treated by a licensed rehabilitator during the day, fed in the afternoon, and taken back to the reunion site for evening.

Occasionally older cubs are found during the daytime. If they appear healthy, just displaced, it may be that they were exploring and did not make it home before sun up. All they might need is a place to stay safe until nightfall when they will reunite with their mother. If safe and practical, a healthy stray may be provided with a cardboard "hut" with soft bedding inside. The rescuers must make sure to keep all people and domestic animals away while they keep vigil until the animal takes off at twilight or is recovered by the mother and other siblings.

Nursing wild canids, like coyotes, have been known to take in and foster very young unrelated pups left outside their den, but not always. Bobcat kittens have also been successfully reunited. When reuniting larger mammals, the babies should be returned to where they were found. They can be housed in a carrier that will keep them sheltered from the elements but through which the mother can see and smell them. They may need to have supplemental warmth if it is cold out, or if they are very young. If possible, the container should be rigged so that the baby can be allowed out when the parent nears. Recordings of the young vocalizing may facilitate the reunion. Rescuers must keep watch from far off. However, if the mother senses humans are nearby, she will be hesitant to approach.

Nutritional support

Diets, formulas, and feeding amounts vary between species; carers can do much more harm than good by giving an animal the wrong food. Providing food also borders on practicing wildlife rehabilitation. Like any advanced aid, this should be performed under the direction of a professional wildlife rehabilitator or wildlife veterinarian. That said, young animals, especially neonates, have specific nutritional needs that must be met for normal development; missed meals can be detrimental.

When a young animal is first acquired, generally, the first few feedings will be a hydrating solution, so it is not likely that rescue personnel will need to worry about nutritional support. However, there may be times when they will need to know what to do in an emergency situation. The following information is offered for those situations where rescuers are not able to receive direction from a wildlife rehabilitator and where missed feedings will place the animal's life in jeopardy.

As a rule, an animal should be warm and sufficiently hydrated before being offered solid food or full-strength formula. When introducing formula to mammals, customarily, the first two feedings are of the hydrating solution only. The subsequent two feedings are often a 50/50 mix of the formula and the hydrating solution. This helps hydrate the animal and allows its body to adjust to the new diet.

In extreme emergency situations, some mammal species can be given a commercially sold formula designed for puppies and kittens, as long as it is appropriate. Kitten milk replacement is suitable for bobcat, raccoon, skunk, and porcupine. Puppy formula, such as Esbilac, is suitable for rodents, fox, coyote, opossum, and mink. Deer and lagomorphs are highly prone to digestive problems. Rescuers should seek the advice of a professional wildlife rehabilitator before providing anything but rehydrating solution to these delicate species.

The amount of formula to feed is based on the animal's weight in grams multiplied by 5%. This equals the number of milliliters the animal can be offered each feeding.

The warmed formula can be offered in a syringe or a pet baby bottle with an appropriately sized nipple. To reduce the chance of it aspirating, the baby should be offered the formula while it lies on its stomach, not on its back.

Baby opossums are the exception. They do not nurse like other mammals. Typically, these marsupials are fed through intubation, or tube feeding. A catheter is used, most often a number $3\frac{1}{2}$ or 5. The distance to the animal's stomach is measured by placing the catheter along the length of the baby opossum's body from its mouth to about the middle of its body. A piece of tape can be used to mark the catheter. The catheter is then attached to a syringe filled with the calculated amount of warmed formula. The air bubbles should be plunged out of the syringe before it is inserted. To insert the tube the carer should moisten it with some of the formula. While the animal is on its stomach, the tube can be gently offered into the right side of the animal's mouth. The opossum will open its mouth and the carer can slowly guide the tube down the throat until the marker reaches the mouth. If there is resistance, the carer should pull back out and try again. The formula should be delivered slowly, watching the nose for bubbles or drops of milk, which indicate a serious problem. If this occurs the catheter should be removed, the excess formula wiped away or the nose cleared with gentle suction from a syringe.

The same general rules apply for birds: they must be warm, housed near 90–95° F (32–35° C), and well hydrated. That does not necessarily mean that feedings are withheld until they are fully hydrated, it just means that hydration must be addressed concurrently.

To decide an adequate emergency diet for a baby bird, one must determine what species it is. A temporary emergency diet for insectivores and omnivorous species would include very well soaked (in water) high-quality, high protein poultry-based kitten kibble with added whites from a hardboiled egg, earthworms, mealworms or crickets. Because baby seedeaters are often fed insects as they are developing they too can be given thoroughly soaked kitten kibble in an emergency. On a temporary basis, doves and pigeons may be gavage-fed a commercial formula, Kaytee Exact. Raptors will require raw food as close as possible to what they would receive in the wild, such as a cut up dead mouse or chick. Ducks and geese can be offered starter mash and young quail or turkey can be offered game bird starter. Again, these are very basic emergency diets offered for rescuers so that a baby bird does not go too long without sustenance.

Developing babies require almost constant feedings throughout their natural photoperiod. Hatchlings are fed every 15–20 minutes; nestlings are usually tended to every 30 minutes; young fledglings are fed about every 30–45 minutes and older fledglings up to an hour between feedings. As a rule, if it is begging, it is hungry.

The delivery method is extremely important, so that the bird's feathers stay clean and dry and the bird does not aspirate. Most songbirds will gape, opening their mouths to be fed. Using a very small, soft paintbrush or blunt-tipped tweezers, a little bird can be offered a meal. In general, the dab of food should be delivered into the mouth and more often than not the baby will do the rest. Depending on their age, raptors can be offered small bits of meat or fully intact rodents. Precocial chicks, such ducklings, can be offered a shallow tray of food along with a shallow container of water, taking care they do not get soaking wet.

18 Offering public service

This chapter is written to and for those who are dabbling with the idea of starting up their very own wildlife search and rescue operation. The following information is offered not so much as direction on how to go about starting your own organization, but more as an example of what has proven successful and what can be accomplished.

As with any business venture, its success will be based, largely, on need. Is there a need for a dedicated wildlife search and rescue operation in your community? What services already exist? How are wildlife casualties responded to, if at all? Where are injured wild animals taken for rehabilitation, or are they destroyed? These are the questions that need to be answered before going any further.

Perhaps a wildlife rehabilitation program exists in your community. Talk with them first. Find out if they offer the type of services you're interested in providing. Perhaps there is a way of joining efforts.

The next thing to consider is how your program will be supported. Will you be able to generate enough interest and support from your community? Will you be able to attract enough volunteers?

Moving forward in establishing a wildlife search and rescue program will involve drafting a plan and starting a list of things that will need to be accomplished. This is where the idea starts to take shape, at least on paper.

One of the things to consider is the name of your program. Giving it a title early on can be beneficial. The name should be easy for people to remember.

In organizing your thoughts about how the program will function, consider the geographic service area – what communities you will serve. You will also want to decide on the types of animals you are willing to assist – if there are any species you will not or cannot respond to – what supplies and equipment you will need, how you will publicize your services, and how you will generate funds.

One of the next steps will involve looking into the legalities of what you want to do. What licenses or permits will you need? Remember to think local, state, and federal.

At some point, as your idea takes shape, you will want to consider incorporating and forming a legal entity, and deciding whether it is going to be a nonprofit or not. Another option to filing for your own nonprofit is to find a fiscal sponsor under which your program can receive its a charitable status.

One of the most critical components of a wildlife search and rescue program is its phone line. The phone is the lifeline to the injured, ill, and orphaned. How calls are received and how they are answered is paramount. Will callers ever hear a busy signal or have to wait for their call to be answered? Will they hear a recording, and if so, will they be compelled to leave their name and number? There are many options to set up an effective yet affordable phone system and some definite DOs and DON'Ts.

As for choosing a number for the wildlife search and rescue program, it should be easy to remember. When it comes to choosing the actual phone system, most landlines will have the option for additional voicemail "boxes" where callers can, for example, leave an emergency related message in one box, or leave a message about volunteering in another.

Another option is to have the emergency call-in number hosted offsite, contracting with a voice mail answering service. The call is routed through their establishment and answered with your own custom menu or by their live operators. This type of system can offer prerecorded tutorials and information or be programmed to route calls, immediately transferring callers to the on-duty Call Taker.

A paging service is also an excellent offsite option. Pagers can be standalone or linked to a simple messaging service. When callers dial the dedicated number, they receive a custom greeting that instructs them to leave a message. When they hang up, the system sends out an alert. Within minutes, easily, the caller can be speaking directly with a live person. Pagers are often preferred as they can be handed off as personnel begin and end shifts. Paging services now offer the option of having messages delivered to Email addresses as WAV files.

More sophisticated, yet still affordable, are computer-based phone systems that employ interactive voice and touch-tone response (IVR) to guide users to specific tutorials. Depending on how a caller responds, verbally or through the keypad, the system can guides them to information pertinent to their specific situation, such as the type of animal they have found. This type of system has been found extremely beneficial to wildlife rescue programs in reducing the number of man-hours spent on non-emergency calls, while providing a valuable service to the community. An example of this is a tutorial that provides information on healthy ducks and ducklings, succinctly delivering the "dos and don'ts" with an option for reaching the Call Taker if it really is an emergency.

The phone is the lifeline to a wild animal in trouble; it is also where the organization makes its first impression on a personal level – connecting with the caller. Each emergency call must be taken seriously and given the attention necessary. With the life of a wild animal at stake, those who answer calls from the public must have the skills to make callers feel as though they did the right thing by phoning. If they feel uncomfortable with what they are being told, or the manner in which they are being addressed, they may stop listening altogether and take matters into their own hands. Good public relations are essential.

Public relations and the art of shapeshifting

The most important role in a wildlife search and rescue program may very well be that of the Call Taker. Through each caller they hold the lifeline to the reportedly disabled animal. It is through the Call Taker's ability to develop and maintain contact that they are able to help the animal. To do so, however, they are often confronted with the extremes of human emotion – panic, frustration, and even anger. There are tricks to working the phone line, and there are certain personalities that will be suited for the role, and others that will not.

The momentous responsibilities and daunting tasks of a wildlife emergency Call Taker are best suited for a person with tremendous patience, tolerance, excellent communication and interpersonal skills, and the ability to maintain professionalism under extreme pressure. They must be capable of being authoritative yet polite, courteous, and able to convey compassion. They must be well informed and versed in the natural history of the species endemic to the region and have a solid understanding of the organization's policies, its role within the community, and the laws and regulations that govern the region's wildlife. They should also have an affinity for solving riddles, and the knack of being a good detective. Most importantly, though, they must have the skills to build a favorable relationship with the stranger on the other end of the line.

To develop their connection with the reporting party (RP), a Call Taker will want to build upon their common interest – the animal, and the fact that they cared enough to call. The key to building and maintaining a link with the caller is through trust. To build trust, the Call Taker will want to develop rapport.

Borrowing from the teachings of neurolinguistic programming, Call Takers might start by making the person on the other end of the line feel more at ease by adopting their rhythm of breathing, or their tempo. The Call Taker should try to modify their speech and delivery to mirror the caller's, embodying a character – shapeshifting, if you will.

Another technique is to repeat a portion of what the caller has stated, making them feel heard, and helping them backtrack. This helps to draw out important details. The Call Taker can even repeat it back as a question: for example, "So, you have found a baby bird, and you're not sure where its mom is?"

It is important for the Call Taker to give the person the opportunity to tell their story, extracting truths from the finders' conjectures. In allowing the caller to speak freely, the Call Taker will be able to get a feel for the person on the other end of the line. If the person is talkative, it may be best for the Call Taker to give him or her instructions through a story. If they are crisp and factual, they may take direction better if it is delivered through cut and dry facts.

For a Call Taker to deliver information fluently he or she will want to have a collection of resources on hand – a few natural history books, a compendium of answers to common questions, perhaps a set of scripts or algorithms, and maps of the organization's service area.

Contracting with municipalities

Once a wildlife search and rescue program has been formally established and is operating smoothly, its directors might want to consider the possibility of contracting with local municipalities to provide a service. Cities could benefit by having a professional, specially trained and equipped team of experts to rely on for wildlife-related calls from the public. The organization could benefit from increased exposure and the credibility the relationship might lend to its program.

Most cities and counties have dedicated service providers for domestic animal issues. As part of their public service, these agencies may assume wildlife response but may not have proper or adequate training, equipment, or holding facilities appropriate for wild animals.

In 1996, Wildlife Emergency Response (Malibu, CA), also known as The California Wildlife Center, entered into a contract with the City of Malibu to provide wildlife services for $1.00 a year. This official relationship allowed the organization to rescue marine mammals, but it also helped establish it as a primary responder. All wildlife-related calls from the public were directed to the organization's hotline. By being named first responder to wildlife emergencies, the organization was able to control the quality of care animals received, saving wild lives that would surely have been lost, returning babies to their families that would have otherwise been orphaned, and providing the community with a valuable expert resource. A sample of that contract can be found in Appendix 6.

APPENDIX 1

Ready packs

Below are suggested items that have proven useful on wildlife rescue missions.

Twenty-four hour pack

Identification
Permits (if necessary)
Medical alert card
Flashlight
Headlamp
Hand sanitizer
Antimicrobial wipes
First aid kit
Sun protection
Bandana
Eye protection
Insect repellant
Exam gloves
Light duty leather gloves
Sharpies
Plastic bags (various sizes)
Cell phone
All-weather writing tablet
Pencils/pens
Emergency phone numbers
Drinking water (I gallon or more)
Electrolytes (packet)
Protein bars
Change of clothes
Extra pair of socks
Extra footwear
Hair tie (for long hair)
Personal hygiene items
Medication (if any)

Toolbox

Multipurpose tool
Good pair of wire cutters
Pliers
Needle-nose pliers
Scissors
Hammer
Shovel

Search and rescue

Binoculars
Maps
Tide charts (if applicable)
Handheld GPS
Two-way radios
Rescue whistle
Appropriate PPE
Heavy duty gloves
Camera
Vehicle placard (if applicable)
Large towels
Sheets
Pillow cases
Mesh dive bag(s)
Cardboard carriers

Wildlife Search and Rescue: A Guide for First Responders, First Edition. Rebecca Dmytryk.
© 2012 John Wiley & Sons, Ltd. Published 2012 by John Wiley & Sons, Ltd.

Plastic carriers
Animal crates
Crate hardware
Nets (various sizes)
Newspaper
Livestock marker
Bag of rags
Garden netting (roll or pack)
Disinfecting solution
Herbal flea repellant sachets
Stretcher or litter

Hell box

Cable ties
Duct tape
Matches
Sewing kit
Super glue
Safety pins
Eyedropper
Twine
Tarpaulin(s)

APPENDIX

2

Wildlife observation form

Incident: _____ Date: _____ Segment Name: _____ State: _____ County: _____
Terrain: _____ Portion covered (start/end GPS locations) _____
Time start: _____ Time end: _____ Tide: _____ Weather: _____ Temp: _____ Wind direction: _____ Beaufort scale: _____
Survey mode: Scope Foot ATV Boat Air Vehicle Other: _____ Survey team: _____
Oil present: Y N   Describe _____ Other: _____

Species	No.	Age	Location	Condition	Time	North coordinates	West coordinates	Collected	Other

APPENDIX 3
Wildlife trauma equipment and supplies

These are items that have proven useful for emergency care of wildlife casualties in the field. These suggested items are specific to wildlife first aid and should be considered in addition to rescue equipment and ready packs. Some of these tools require special training, and skill to use. The administration of advanced medical care will be provided under the direction of a licensed veterinarian.

Equipment

Exam gloves
Nitrile gloves
Respirator mask
Medical scissors
Stethoscope
Laryngoscope
Cuffed endotracheal tubes
Ambu bag
Artery forceps
Tweezers
Thermometer
Pen light
Tongue depressors
Eye dropper

Bandaging

Sterile swabs
Cotton balls
Rolled cotton
Cloth tape
Non-stick pads
Gauze dressing pads
Rolled gauze
Waterproof cloth tape
Vet wrap
Band-Aids
Non-sterile 4 × 4 gauze pads

Solutions, ointments, gels

Isopropyl alcohol
Hydrogen peroxide
Sterile saline solution
Betadine solution
Ophthalmic ointment
Wound honey
K-Y lubricating jelly
Hemostatic powder

Thermoregulation

Space blanket
Heating pad (12-volt)
Hot pack
Dry rice beanie
Spritzer
Cold pack

Fluid therapy

Electrolyte powder/solution
Water

Wildlife Search and Rescue: A Guide for First Responders, First Edition. Rebecca Dmytryk.
© 2012 John Wiley & Sons, Ltd. Published 2012 by John Wiley & Sons, Ltd.

Lactated Ringer's
Normosol
Intravenous drip set
French catheters
Syringes
Feeding tube chart
Needles

Other

First aid protocol
Sharps container

Biohazard bag
Ziploc bags
Specimen canisters
Tape measure
Digital scale
Calculator
Glucose test strips

APPENDIX 4
Instructions for tying nooses

Tying monofilament line nooses

Step 1:
Create single loop: Loop A.

Step 2:
Cross Loop A over itself to create Loop B. Move Loop A behind.

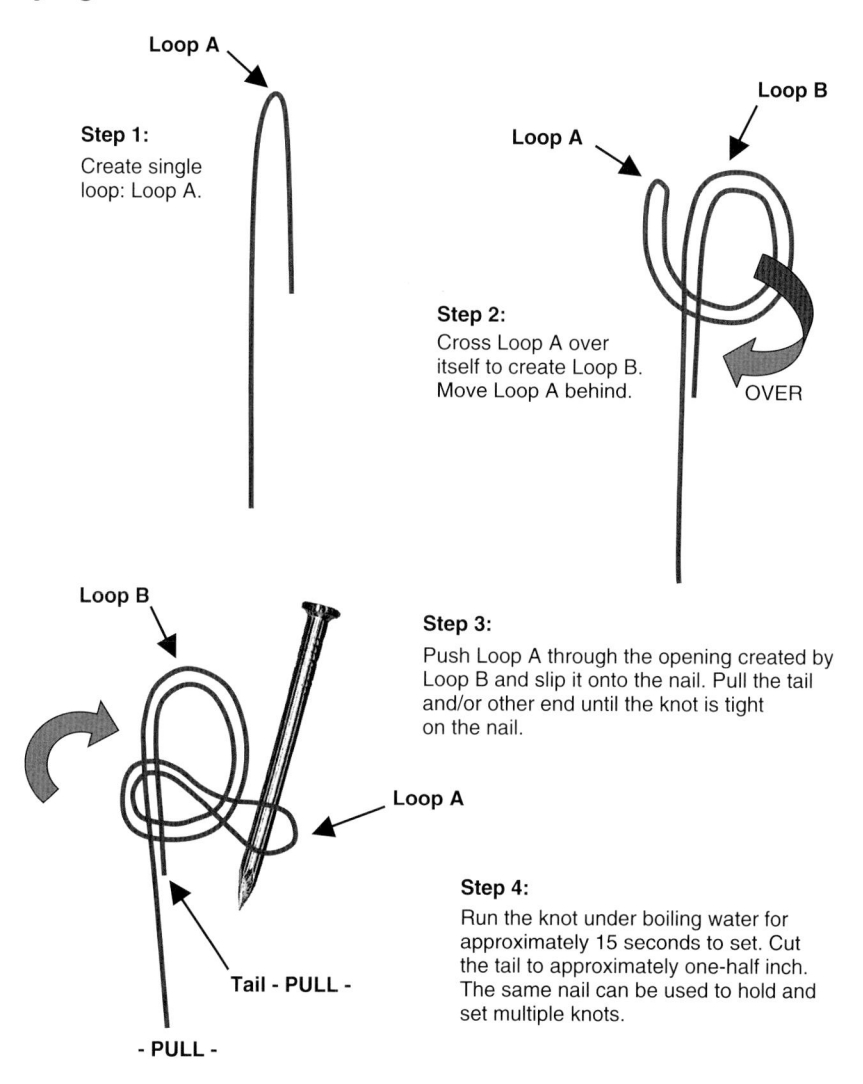

Step 3:
Push Loop A through the opening created by Loop B and slip it onto the nail. Pull the tail and/or other end until the knot is tight on the nail.

Step 4:
Run the knot under boiling water for approximately 15 seconds to set. Cut the tail to approximately one-half inch. The same nail can be used to hold and set multiple knots.

Wildlife Search and Rescue: A Guide for First Responders, First Edition. Rebecca Dmytryk.
© 2012 John Wiley & Sons, Ltd. Published 2012 by John Wiley & Sons, Ltd.

Tying nooses to welded wire

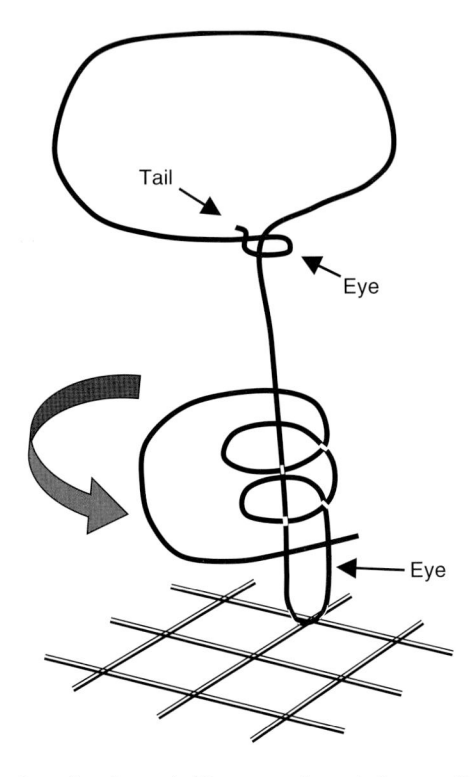

Create the noose by threading the end of the noose through the eye. To tie it to the wire, take the long end and thread it under an intersection of wire. Wrap it around the base of the noose two or three times before threading it through the bottom eye, then pull to tighten. Check to see if the noose stands perpendicular. Adjust as needed until it stands up. Apply glue to the knot on welded wire. Cut the remainder of the knot's tail to one-half inch.

APPENDIX 5

Barn owl box plans and instructions

The owl nest box should be mounted on a secure pole so that it is at least 12 feet high and away from disturbances. In some cases it can be secured to a tree or a building. Proper placement is very important; the box entrance should face away from prevailing winds and must not be blocked by structures, as the owls tend to swoop in.

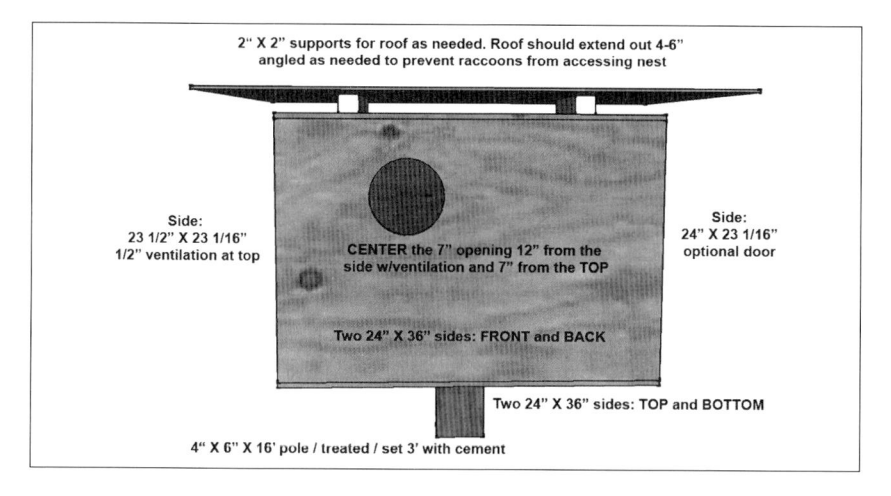

```
2" X 2" supports for roof as needed. Roof should extend out 4-6"
angled as needed to prevent raccoons from accessing nest
```

Side:
23 1/2" X 23 1/16"
1/2" ventilation at top

CENTER the 7" opening 12" from the side w/ventilation and 7" from the TOP

Side:
24" X 23 1/16"
optional door

Two 24" X 36" sides: FRONT and BACK

Two 24" X 36" sides: TOP and BOTTOM

4" X 6" X 16' pole / treated / set 3' with cement

Here is the cut list for one 4 by 8 feet sheet of 15/32 inch (or 1/2 inch) plywood. The shade top must be cut from a separate piece of lumber and can be thinner.

Wildlife Search and Rescue: A Guide for First Responders, First Edition. Rebecca Dmytryk.
© 2012 John Wiley & Sons, Ltd. Published 2012 by John Wiley & Sons, Ltd.

Appendix 5

The pieces are assembled using pieces of 2 by 2 inch lumber and exterior wood screws called grabbers. The *bottom* is the last piece to be attached - after the box has been mounted on the pole and the shade top has been secured. For the pole, a 4 inch by 6 inch by 16 feet pressured treated piece of lumber is recommended.

Note: the two *ends* are set between the *front* and *back* pieces, that measure are 24 by 36 inches, and the *front* and *back* sections are set between the *top* and *bottom*.

Notice the placement of a cleat below the hole to help the growing birds reach the exit when they are ready. It is extremely important that any steps are placed no less than 8 inches from the opening. They can be made from 1 inch by 2 inch pieces of lumber. Make sure the screws point outward!

A few 1/4 inch holes can be drilled through the floor for drainage.

The unit can be protected with clear wood sealant or primed and painted with exterior acrylic paint. Make sure it is thoroughly dry before it is mounted.

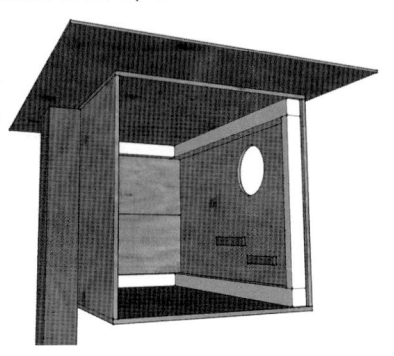

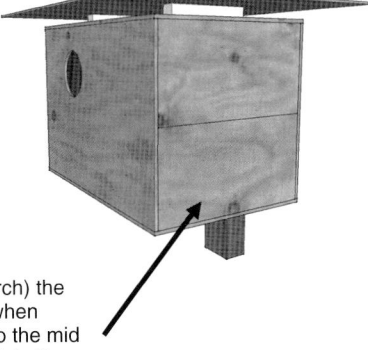

If future access is desired (for cleaning or research) the door should be placed on this end. Make sure, when mounting the hinges to the base, and the latch to the mid section, that the screws do not penetrate through to the living space where the owls might become injured.

APPENDIX 6

Sample contract

This is a sample contract between a government entity and a nonprofit organization for wildlife rescue services. If this contract were to be utilized, the word "municipality," below, would name the government entity and its abbreviated name would replace ("City"), removing parentheses and quote marks as needed.

Agreement for professional services

THIS AGREEMENT is made and entered into on this _____ (day) _____ day of _____ (month) _____ , _____ (year) _____ , by and between (municipality) (hereinafter referred to as ("City"), and _____ (the name of the organization) _____ , hereinafter referred to as Contractor.

The ("City") and Contractor agree as follows:

Recitals

A. The ("City") requires wildlife search and rescue services for sick, injured, and orphaned wildlife within _____ (region, i.e. city limits) _____ , ancillary to those services currently provided by existing government agencies, including but not limited to, _____ (other service providers if any) _____ .

B. The ("City") does not have the personnel able and/or available to perform the services required under this Agreement.

C. Contractor has presented to ("City") that it is possessed of the experience, skills, and abilities to furnish such wildlife services for ("City") and it desires to contract with ("City") to provide the same on ("City's") behalf.

D. The ("City") desires to contract with Contractor to perform the services as described in Exhibit A of this Agreement.

Wildlife Search and Rescue: A Guide for First Responders, First Edition. Rebecca Dmytryk.
© 2012 John Wiley & Sons, Ltd. Published 2012 by John Wiley & Sons, Ltd.

In consideration of the mutual terms, conditions and covenants hereinafter set forth, ("City") and Contractor agree as follows:

NOW, THEREFORE, the ("City") and Contractor agree as follows:

1.0 Scope of services

Contractor agrees to provide, on an as needed basis, wildlife search and rescue services as more fully detailed in SCOPE OF SERVICES attached hereto and incorporated herein as Exhibit A. Contractor shall devote such time, energies, and attentions as required to perform this Agreement in a professional and competent manner. For the purpose of performing said duties and for the purpose of giving official status to the performance thereof, all Contractor personnel engaged in performing such services shall be deemed to be an official of ("City") while performing such services. The Scope of Services may be amended from time to time by way of written directive from the ("City").

1.1 Obligations of contractor

Such services shall be rendered by following the procedures attached hereto and incorporated herein as Exhibit B. In performance of these services, Contractor shall comply with the following:

(a) Provide ("City") with an annual report indicating all names and addresses of personnel working in a paid employment or volunteer capacity for Contractor;

(b) Provide ("City") with annual reports of service activities, including but not limited to, the number of wildlife rescue-related phone calls received, the number of wildlife search and rescue missions performed, and the number and types of animals encountered. Reports shall be submitted on _____ (date) _____ of each year;

(c) Provide ("City") with an annual financial report of all agency revenues and expenditures.

2.0 Term of agreement

This Agreement shall be for a term of _____ (i.e. one year) _____ commencing on (date) and terminating (date). Either party may, with or without cause, terminate this Agreement by giving fifteen (15) days written notice to the other party.

3.0 Compensation

The ("City") shall pay Contractor, and Contractor shall accept from ("City"), total compensation for its professional services rendered and costs incurred pursuant to this Agreement for the sum of _____ (e.g. one dollar/$1.00) _____ during the entire period of the Agreement. No additional compensation shall be paid for any other expenses incurred, unless first approved by the _____ (i.e. City Manager) _____ , or his/her designee.

4.0 General terms and conditions

4.1 Non-assignability

Contractor shall not assign or transfer any interest in this Agreement without express written consent of the ("City").

4.2 Compliance with law and non-discrimination

Contractor shall not discriminate as to race, creed, skin color, national origin, or sexual orientation in the performance of its services and duties pursuant to this Agreement, and will comply will all applicable laws, ordinances, and codes of Federal, State, County, and City governments, including but not limited to (e.g. the US Fish and Wildlife Service). Contractor is authorized to enter _____ (e.g. city beaches) _____ with motor vehicles for the sole and express purpose of carrying out its responsibilities under this Agreement.

4.3 Insurance

Contractor shall submit to the ("City") certificates indicating compliance with the following minimum insurance requirements no less than one (1) day prior to beginning performance under this Agreement:

(a) Workers Compensation Insurance as required by law. Contractor shall require all subcontractors similarly to provide such compensation insurance for their respective employees.

(b) Liability insurance in the amount no less than $1,000,000.00. Such policy shall:

1. Be issued by a financially responsible insurance company admitted and authorized to do business in the State of _____ (state) _____ or which is approved in writing by ("City").
2. Name ("City") and its officers and employees as additional insured.
3. Cover the operations of Contractor pursuant to the terms of this Agreement.
4. Hired and Non-Owned vehicle insurance if applicable.

4.4 Indemnification

Contractor shall not have any authority to bind or incur any debt on behalf of ("City"). ("City") shall not be responsible for direct payment of any salaries, wages, or other compensation to Contractor personnel. Contractor agrees to hold harmless, indemnify and defend the ("City"), its employees, agents, and affiliates, for any and all loss or liability of any nature whatsoever arising out of or in any way connected with Contractor's performance of this Agreement.

4.5 Independent contractor

This Agreement is by and between the ("City") and Contractor and is not intended, and shall not be construed, to create the relationship of agency, servant, employee, partnership, joint venture, or association, as between the ("City") and Contractor.

4.5.1.

Contractor shall be an independent contractor, and shall have no power to incur any debt or obligation for or on behalf of the ("City"). Neither the ("City") nor any of its officers or employees shall have any control over the conduct of Contractor, or any of Contractor's employees, except as herein set forth, and Contractor expressly warrants not to, at any time or in any manner, represent that it, or any of its agents, servants, or employees are in any manner employees of the ("City"), it being distinctly understood that Contractor is and shall at all times remain to the ("City") a wholly independent contractor and Contractor's obligations to the ("City") are solely such as are prescribed by this Agreement.

4.6 Legal construction

(a) This Agreement is made and entered into in the State of _____ (state) _____ and shall in all respects be interpreted, enforced and governed under the laws of the State of _____ (state) _____ .

(b) This Agreement shall be construed without regard to the identity of the persons who drafted its various provisions. Each and every provision of this Agreement shall be construed as though each of the parties participated equally in the drafting of same, and any rule of construction that a document is to be construed against the drafting party shall not be applicable to this Agreement.

(c) The article and section, captions and headings herein have been inserted for convenience only, and shall not be considered or referred to in resolving questions of interpretation or construction.

(d) Whenever in this Agreement the context may so require, the masculine gender shall be deemed to refer to and include the feminine and neuter, and the singular shall refer to and include the plural.

4.6.1 Partial invalidity.

If any provision in this Agreement is held by a court of competent jurisdiction to be invalid, void, or unenforceable, the remaining provisions will nevertheless continue in full force without being impaired or invalidated in any way.

4.6.2 Attorneys' fees.

The parties hereto acknowledge and agree that each will bear his or its own costs, expenses and attorneys' fees arising out of and/or connected with the negotiation, drafting and execution of the Agreement, and all matters arising out of or connected therewith except that, in the event any action is brought by any party hereto to enforce this Agreement, the prevailing party in such action shall be entitled to reasonable attorneys' fees and costs in addition to all other relief to which that party or those parties may be entitled.

4.6.3 Entire agreement.

This Agreement constitutes the whole agreement between the ("City") and Contractor, and neither party has made any representations to the other except as expressly contained herein. Neither party, in executing or performing this Agreement, is relying upon any statement or information not contained in this Agreement. Any changes or modifications to this Agreement must be made in writing appropriately executed by both the ("City") and Contractor.

This Agreement is executed on this _____ (day) _____ of _____ (month) _____ , _____ (year) _____ , at _____ (location _____ , _____ (state) _____ .

(Name of city)

_____

(Name of officer, Title)

_____

(Name of organization)

_____

By: (Name of representative)

Title: (Title)

Exhibit A: Scope of services

This is where the organization can spell out what services it intends to provide under the contract. The following is merely an example

Provide 24-hour assistance for emergencies involving native wildlife, including but not limited to the following:

(a) 24-hour helpline with a way for callers to reach a live Call Taker.

(b) Provide field response, as needed, including but not limited to:

- Wildlife search and rescue services.
- Initial aid to wildlife casualties, as needed.
- Transport of wildlife casualties to appropriate facilities, as needed.
- Assist in reuniting/renesting healthy young, as needed.

Exhibit B: Policies and procedures

This is where the organization describes how they it carry out the services it intends to provide. This is also where it can detail its policies, such as following a certain code of conduct, provided here as an example.

In its course of duty and in carrying out wildlife rescue operations, _____ (Name of organization) _____ , its agents, officers, directors, volunteers, staff, and employees are required to act under the following principles:

1 In the course of responding to and rescuing a wild animal, human safety will take precedence; no human life will knowingly be placed in jeopardy to save an animal.
2 Wildlife rescues will be performed in accordance with all applicable laws.
3 Every effort will be made to reduce and minimize pain, stress, and suffering of an animal during capture, handling, and transport.
4 Documentation of the animal or the rescue attempt, whether through photography, video, or audio recordings, will be done in such a way that will not subject the animal to any additional or unnecessary handling or delay its care.
5 Capture strategies, handling techniques, tools, or methods of confinement that pose a significant risk to an animal's life shall be avoided.
6 The administration of advanced medical care will be provided under the direction of a licensed veterinarian.
7 Terminally injured or ill animals will receive the appropriate immediate care governed by the organization's written euthanasia policy.
8 When presented with an otherwise healthy but displaced young wild animals still requiring parental care or guidance, all options will be considered and every effort will be made to keep them with wild parents.
9 When returning animals to the wild, every effort will be made to return them to their home territory.

Further reading

American Veterinary Medical Association (2007) *AVMA Guidelines on Euthanasia*. American Veterinary Medical Association (AVMA), Schaumburg, IL.

Barnett, J., Knight, A. and Stevens, M. (2004) *Marine Mammal Medic Handbook*, 4th edn. British Divers Marine Life Rescue.

Bartos, R., Olsen, P. and Olsen J. (1989) The Bartos trap: a new raptor trap. *Journal of Raptor Research* 23: 117–20.

Bird, D. M. and Bildstein, K. L., eds (2008) *Raptor Research and Management Techniques*. Hancock House Publishers, Canada.

Cooper, D. C., ed. (2005) *Fundamentals of Search and Rescue*. NASAR and Jones and Bartlett Publishers, Sudbury, MA.

Deal, T. (2010) *Beyond Initial Response ICS*, 2nd edn. AuthorHouse, Bloomington, IN.

Dickens, M. J., Delehanty, D. J. and Romero, L. M. (2009) Stress and translocation: changes in the stress physiology of translocated birds. *Proceedings of the Royal Society B* 276: 2051–6.

Dickens, M. J., Delehanty, D. J. and Romero, L. M (2010) Stress: an inevitable component of translocation procedures. *Biological Conservation* 143(6): 1329–41.

Eilertsen, N. and MacLeod, A. (2001) *Flying Chance*. East Valley Wildlife, Phoenix, AZ.

Fowler, M. E. (1995) *Restraint and Handling of Wild and Domestic Animals*, 2nd edn. Iowa State University Press, Ames, IA.

Friend, M. and Franson, J. C. (1999) *Field Manual of Wildlife Diseases. General Field Procedures and Disease of Birds. Information and Technology Report 1999–2001*. US Geological Survey, Madison, WI.

Gage, L. J. and Duerr, R. S. (2007) *Hand-Rearing Birds*. Blackwell Publishing, Ames, IA.

Hadfield, C. A. and Whitaker, B. R. (2005) Amphibian emergency medicine and care. *Seminars in Avian and Exotic Pet Medicine* 14: 79–89.

Hanger, J. and Tribe, A. (2005) Management of critically ill wildlife: the reality and practice of wildlife euthanasia. In *Proceedings of the 3rd National Conference on Wildlife Rehabilitation*. Gold Coast, Queensland.

International Wildlife Rehabilitation Council (2005) *Basic Wildlife Rehabilitation 1AB*, 6th edn. International Wildlife Rehabilitation Council (IWRC), Suisun, CA.

Johnson-Delaney, C. A. (2010) Critical care of reptiles. In *Preserving the Wild in Wildlife*. Symposium conducted at the meeting of the National Wildlife Rehabilitators Association conference, Seattle, WA, March.

Johnson-Delaney, C. A. (2010) Critical care of amphibians. In *Preserving the Wild in Wildlife*. Symposium conducted at the meeting of the National Wildlife Rehabilitators Association conference, Seattle, WA, March.

Lindstrom, A., Klaassen, M. and Lanctot, R. (2005) The foldable "Ottenby" walk-in trap: a handy and efficient wader trap for expedition conditions. *Wader Study Group Bulletin* 107: 50–3.

McCleery, R. A., Lopez, R. R. and Silvy, N. J. (2007) An improved method for handling squirrels and similar-sized mammals. *Wildlife Biology in Practice* 3: 39–42.

Miller, A. G. (2007) *Calls of the Wild*. Alabama Wildlife Center, AL.

Miller, E. A., ed. (2000) *Minimum Standards for Wildlife Rehabilitation*, 3rd edn. National Wildlife Rehabilitators Association, St Cloud, MN.

Moore, A. T. and Joosten, S. (2002) *Principles of Wildlife Rehabilitation*, 2nd edn. National Wildlife Rehabilitators Association (NWRA), St Cloud, MN.

Shimmel, L. (2010) What good is wildlife rehabilitation? Retrieved from http://www.eraptors.org/whatgood.htm.

Stocker, L. (2005) *Practical Wildlife Care*, 2nd edn. Wiley-Blackwell, Oxford.

Tseng, F. S. and Mitchell, M. A., eds (2007) *Topics in Wildlife Medicine: Emergency and Critical Care*. National Wildlife Rehabilitators Association, St Cloud, MN.

US Department of Transportation and US Coast Guard (USCG) (1998) *Team Coordination Training Student Guide*. USCG Publication, Washington, DC.

Van Doninck, H. (2003) *Initial Wildlife Care*. International Wildlife Rehabilitation Council (IWRC), Suisun, CA.

Williams, E. S. and Barker, I. K. (2001) *Infectious Diseases of Wild Animals*, 3rd edn. Blackwell Publishing, Ames, IA.

Index

Accidental Death and Disability, 3
Akashiwo sanguinea, *see* "sea slime", 113
alarm response, *see* stress
alcids, capture methods, 98
altricial, birds, 187, *187*
American coot, *see* rails
amphibians
    capture, 145
    field euthanasia, 180
    fluid therapy, 169
    restraint, 145
animal handling, *see* restraint techniques
anteater, *see* xenathrans
anthrax, *see* zoonotic diseases
armadillo, *see* xenathrans
aspergillosis, *see* zoonotic diseases
avian casualties
    arrows, bolts, 105–6
    avian botulism, 110
    domoic acid, 111–13
    ducklings in pool, 101–2, 101
    entangled in fishing line, 104–5
    glue traps, 109–10
    lead poisoning, 111
    rodenticides, 105
    "sea slime", 113
    trapped in buildings, 98–9
    window strikes, 100

badger, capture and handling, 138–9
    special caging, 144
bal-chatri, 76–8, *77*
ball bandage, birds, 176
bandaging, *see* field dressing
Bartos trap, 76, *77*

bats
    handling, 134–5, *135*
    temporary housing, 143–4
Baylisascaris procyonis, 19
beaver, *see* rodents
bird lice, 21
bleeding, 161
bobcats, *see* felids
body wrap, birds, 176
brancher, *see* fledgling
briefings, 41
brown pelican, capture methods, 95–7
    handling method, 96–7, *96*, *97*
brucellosis, *see* zoonotic diseases
buddy system, 26
bumper wounds, 129

cage traps, 68
Call Taker, *see* rescue personnel
canids
    capture, 133
    restraint, 139–40, *140*
    temporary housing, 144
capillary refill time (CRT), 161
capture
    fundamentals of, 45–51
    using stealth, 48–50
capture myopathy, 38–9
carriers, *see* temporary housing
catchpole, 70, 139
cetaceans, rescue, 156–7
chlamydiosis, *see* zoonotic diseases
cigarette hold, *see* martini hold, 120, *121*
circulatory system, 160–161
clothing, 33–4

Wildlife Search and Rescue: A Guide for First Responders, First Edition. Rebecca Dmytryk.
© 2012 John Wiley & Sons, Ltd. Published 2012 by John Wiley & Sons, Ltd.

Collarum, 133
contageous ecthyma, *see* orf
cormorants
    capture methods, 97
    restraint, 122
    temporary housing, 126
cortisol, released during stress response, 36
coyote, *see* canids
cryptococcosis, *see* zoonotic diseases

debrief, definition of, 41
    oil spill response, 116
deer
    capture and restraint, 134
    temporary confinement, 144
dehydration, 161
    fluid therapy, 162–70
    in rescue personnel, 14
    in neonates, 195
dho-gaza, 66, *67*
distress, *see* stress
dolphins, *see* cetaceans
drive trap, 63, *63*
drop trap, 66–8, *67*
    used on flighted duck, 102

echinococcosis, *see* zoonotic diseases
egrets, *see* waders
ehrlichiosis, *see* zoonotic diseases
Emergency Medical Services (EMS), 4
ermine, *see* mustelids
Erysipelothrix rhusiopathiae, *see* zoonotic diseases
escape route, 45
euthanasia,
    methods, 179–81
    philosophical considerations, 182
exertional myopathy, *see* capture myopathy

felids, restraint, 140
field dressing, 173
figure-eight wing wrap, 174–6, *175*
first aid, 158–76
fishers, *see* mustelids

fish handler's finger, *see* zoonotic diseases
fishing line injuries, *see* bird casualties
fledglings
    reuniting, 187–9, *187*
flight-or-fight response, 35
floating gill nets, 88–9
floating mist nets, 89
floating pen, pinnipeds, 152
fluid therapy, 162–70
fox, *see* canids
fracture stabilization, 173–6
funnel trap, *see* drive trap

gavage, 165–6, *165*
    baby opossum, 198
General Adaptation Syndrome, 36
ghillie suit, 34
giardia, *see* zoonotic diseases
glottis, 165, *165*
glucocorticoid, released during stress response, 36
go bags, *see* ready packs
grebes
    capture methods, 94
    keel sores, 126
    skeletal structure, 127
    special housing, 126–7

hantavirus, *see* zoonotic diseases
hares, *see* lagomorphs
hazards
    definition of, 11
    environmental, 11–12
    equipment related, 13
    hazardous material, protection from, 24–5
    human factor, 12–13
heat injury, *see* hyperthermia
herding
    drive or funnel traps, 22, 64
    understanding pressure, 47–48
herding boards, (also pig boards), 55–7, *56*, 130, 148
    used with open-ended net, 59, 149–50, *149*

herons, see waders
histoplasmosis, see zoonotic diseases
hoop net, (or hand net), 57
   netting birds, 71–5, *72,73, 74*
   netting mammals, 130
   pinnipeds, 148–9
hot wash, see debrief
hotline, see phone systems
human safety, 11–15, 22–7
Humane traps, see cage traps
hummingbirds
   in skylight, 99
   nutritional support, 92–3, *93*
   restraint, 118, *118*
hunting
   in development of conservation laws, 7
   skills used in rescue, 42
hydatidosis, see zoonotic diseases
hyperthermia
   and capture myopathy, 39
   diagnosis, treatment, 171–2
   during restraint, 39
   in rescue personnel, 14
hypothermia, 170–171
   in neonates, 195

Incident Command System (ICS), 40, 117
initial assessment, 158–9, 184, 186
   by the Call Taker, 52
International Bird Rescue, 127, 129

jizz, 43
jug-handling, 43, 146

keel, 126, 127, 160
   "doughnut" cushion, 127, 128
   injury, preventative measures, 126–7

lagomorphs,
   capture, 132
   restraint, 138
land seine, 60–61
larva migrans, see zoonotic diseases
leptospirosis, see zoonotic diseases

loons,
   capture methods, 93–4
   keel sores, 126–7
   skeletal structure, 127
   special housing, 126–7
lures,
   audio, 70
   baiting birds, 71–6
lyme disease, see zoonotic diseases

mange, see zoonotic diseases
Marine Mammal Protection Act, 8–9
marine mammals
   authorization to rescue, 9
   rescue, 146–7
martens, see mustelids
mechanical nets
   land seine, 60–61
   bow net, 61
   whoosh, 62, *62*
medical alert card, 32
mice, see rodents
Migratory Bird Treaty Act, 8
Minimum Standards for Wildlife
   Rehabilitation (MSWR), 9
mink, see mustelids
mist nets, 86
moles, see talpids
murine typhus, see zoonotic diseases
mycoplasma infections, see zoonotic
   diseases

National Marine Fisheries Service (NMFS),
   8
natural history, 1, 38, 42–3
   in reducing stress, 38
   in reuniting, 186
neonates, mammals
   evaluation, 195
nestling
   reuniting, 187–9
net cannon, (also net launcher), see
   projectile-powered nets
net-bottom caging, 127, *128*

net gun, *see* projectile-powered nets
netting, mesh, 57
nutritional support, neonates, 198–9

Occupational Safety and Health administration (OSHA), 24
oil, petroleum, 24
    protective equipment, 25–6, 33
open-ended hoop net
    construction, 58–9, *59*
    use, 17, *59*
    pinniped rescue, 148–50, *149*
open mouth breathing, 172
operational risk management (ORM), 27–32
opossums
    restraint, 137, *138*
orf, *see* zoonotic diseases
otariids, *see* pinnipeds
Ottenby, *see* walk-in trap
otter, river, *see* mustelids

pain guarding, 158
parrots, restraint, 123
passerines
    restraint, 118, *118*, 119
Peg, the plover, 182–3
pelicans, capture, restraint, 97, *97*, 122
    administering oral fluids, 167, *167*
personal protective equipment (PPE), 22–6
pig boards, *see* capture equipment
Pinky, the turkey, 106
pinnipeds
    confinement, 153, 154, *155*
    natural history, 146–7
    rescue, 146–55
    restraint, 152–3, *153*, *154*
    transport, 148, 153–4
pit traps, 85–6
phai trap, 78–9, *78*
phocids, *see* pinnipeds
phone systems, 201
physical exam, 159–61
    young birds, 188, 189

placement, captivity, 182
plague, *see* zoonotic diseases
porcupine, *see* rodents
porpoise, *see* cetaceans
precocial, baby birds, 186, *187*
preferred optimal temperature zone (POTZ), 169
pressure, *see also* herding, 45, 64, 103
    on water, 89
    principles of, 47–8
    ways of reducing, 50–51
    when luring birds, 72
processing
    birds, 124
    chute, 141–2
    land mammals, 141–2, *142*
    marine mammals, 148–9, *149*
projectile-powered nets, 68–9
    net gun, 69
pro-words, use of, 41
psittacosis, *see* zoonotic diseases
public relations
    connecting with finder, 5, 202–3

Q fever, *see* zoonotic diseases

raccoon
    restraint, 138–9
    roundworm(Baylisascaris procyonis), *see* zoonotic diseases
rabbits, *see* lagomorphs
rabies, *see* zoonotic diseases
rails
    American coot, 95
    capture methods, 94
    restraint, 123
raptors
    restraint, *120*, 122, 123, *123*
    use of the abba, 122
rats, *see* rodents
ready packs, 32
re-nesting, *see* reuniting
reptiles
    capture, 57, 145

reptiles (*continued*)
  field euthanasia, 180
  first aid, 160, 162, 164, 169
  fluid therapy, 169–70
  restraint, 145
Rescue Coordinator, *see* rescue personnel
rescue equipment, consolidating, 32–3
rescue operations, planning for, 40–41
rescue personnel
  Call Taker, 4, 52–3, 202–3, 158
  capabilities, 12
  evaluation of, 12–13
  leadership qualities, 40, 41
  Rescue Coordinator, 53
  roles, 52–4
  Team Leader, 53
  Transport Team, 54
  Wildlife Paramedic, 53
restraint techniques
  abba, 122
  head of bird, 120–121, *121*
  game birds, 123–4
  general rules, 39
  martini hold, 120–121, *121*
  popsicle hold, 118, *120*
  ringer's hold, 118, *119*
  using a net, 130, *131*
  waterfowl, 119, *121*
reuniting
  altricial chicks, 187
  fledglings, branchers, 185–9
  mammals, 195–7
  nestlings, 187–9
  precocial chicks, 193–4
  preconditions, 186
  program, 185
rickettsial infections, *see* zoonotic diseases
ringworm, *see* zoonotic diseases
risk
  definition of, 11
  management, 27–32
Robert Jones bandage, 174
rodents
  capture, 131–2
  cone-shaped collar, 137
  fabric handling bag, 137
  handling porcupines, 137
  restraint, 135–7, 136
  rodenticide poisoning, birds, 105

safety briefings, 13, 41
safety equipment, *see* personal protective equipment
safety protocols, general, 26–7
salmonellosis, *see* zoonotic diseases
sea lions, *see* pinnipeds
seals, *see* pinnipeds
seal finger, *see* zoonotic diseases
search, fundamentals, 44–5
  formations, 44, *44*
shags, *see* cormorants
short-term housing
  birds, 124–5
shrews, *see* talpids
site fidelity, 44
  in American coots, 95
situational awareness, 12–13
skunks
  capture, 133
  restraint, 139
  special caging, 133, 144
sloth, *see* xenathrans
snares
  leg snare pole, 84–5
  noose carpets, 76, 79
  single leg snare, 79–80, *80*, 83
  tying nooses, *209*
soft-sided caging, 127
songbird, *see* passerines, restraint, 118–19, *118*, *119*
spotlighting,
  on land, 90–91, *91*
  from a boat, 92, *92*
squirrels, *see* rodents
Star Of Life, 4
The Star Thrower, 5
static charge, bedding, 143
stoats, *see* mustelids

stress, on animals, 34–8
  alarm response, 36
  defined, 35
  distress, 35–6
  in a rehabilitation setting, 37
  minimizing, 38
  physiological response, 36
submersible nets and pens, 89–90, *90*
subcutaneous injections, 168–9
sunburn
  protection, 14
supplemental heat, 170–171
supportive care
  privacy and quiet, 37
swan hook, 85
talpids, handling, 136
temporary housing
  bedding material, mammals, 143
  birds, 125–9
  land mammals, 142–4
throw net, 60
thygmotaxis
  in otariids, 146
  rodents, 132
ticks
  diseases from, 15–16
towel wrap, pinniped rescue, 151
towels, uses for, 55
  material, 143
toxoplasmosis, *see* zoonotic diseases
transporting wildlife, 177–8
tubing, *see* gavage
tularemia, *see* zoonotic diseases

U. S. Fish And Wildlife Service (FWS), 8

veterinary care, birds, under MBTA, 8

waders, 97–8
walk-in traps, 64
  Ottenby, 64, 65
water rescues, 86–90,
  bird onshore, 87, *87*
  on the water, 87–8, *88*

weasels, *see* mustelids
West Nile virus, *see* zoonotic diseases
whale rescue, *see* cetaceans
wildlife
  conservation, 7
  laws, 7–8
Wildlife Paramedic, *see* rescue personnel
wildlife rehabilitation
  definition of, 4
  regulatory agencies, 7–8
  value of, 4, 5
wildlife search and rescue operations
  appointing roles, 12
  code of practice, 10
  definition of, 3
  establishing a program, 200–201
  evaluating team members, 12
  potential for success, 11
  roles within a team, 52–4
  skills, importance of, 1, 2, 3
  standards, 3, 9, 17
  value of, 5
Wildlife Search And Rescue Technician
    (WSART), 1, 2
Wildlife Observation Form, 44–5
wing injuries
  assessment, 158, *159*
  bandaging, 174–6, *175*
wound management, 172–3

Xenarthrans, capture, 132–3

zoonotic diseases
  anthrax, 15
  aspergillosis, 18
  brucellosis, 15
  chlamydiosis, 18
  cryptococcosis, 18
  echinococcosis, 20
  ehrlichiosis, 15–16
  fish handler's finger, erysipelothrix
      infection, 16
  giardia, 20
  hantavirus, 19

zoonotic diseases (*continued*)
  histoplasmosis, 18
  hydatidosis, 20
  larva migrans, 19–20
  leptospirosis, 16
  lyme disease, 16
  mange, sarcoptic, 20
  Murine typhus, 16
  mycoplasma infections, 16
  orf, 19
  plague, 16
  psittacosis, 18
  Q fever, 17
  rabies, 19
  rickettsial infections, 17
  ringworm, 18
  Rocky Mountain spotted fever, 17
  salmonellosis, 18
  seal finger, 16
  toxoplasmosis, 20
  tularemia, 18
  West Nile virus, 19